d Food
Association Group

the **information** store
📞01603 773114
email: tis@ccn.ac.uk

21 DAY LOAN ITEM

Please return <u>on or before</u> the last date stamped above

A fine will be charged for overdue items

CITY COLLEGE NORWICH

CCFRA Tec

210 267

© CCFRA 2004

Campden & Chorleywood Food
Research Association Group

Chipping Campden, Gloucestershire, GL55 6LD UK
Tel: +44 (0) 1386 842000   Fax: +44 (0) 1386 842100
www.campden.co.uk

Information emanating from this company is given after the exercise of all reasonable care
and skill in its compilation, preparation and issue, but is provided without liability in its
application and use.

The information contained in this publication must not be reproduced without permission
from the CCFRA Publications Manager.

Legislation changes frequently. It is essential to confirm that legislation cited in this
publication and current at the time of printing is still in force before acting upon it. Any
mention of specific products, companies or trademarks is for illustrative purposes only and
does not imply endorsement by CCFRA.

© CCFRA 2004
ISBN: 0 905942 69 8
A catalogue record for this book is available from the British Library.

SERIES PREFACE

Food and food production have never had a higher profile, with food-related issues featuring in newspapers or on TV and radio almost every day. At the same time, educational opportunities related to food have never been greater. Food technology is taught in schools, as a subject in its own right, and there is a variety of food-related courses in colleges and universities - from food science and technology through nutrition and dietetics to catering and hospitality management.

Despite this attention, there is widespread misunderstanding of food - about what it is, about where it comes from, about how it is produced, and about its role in our lives. One reason for this, perhaps, is that the consumer has become distanced from the food production system as it has become much more sophisticated in response to the developing market for choice and convenience. Whilst other initiatives are addressing the issue of consumer awareness, feedback from the food industry itself and from the educational sector has highlighted the need for short focused overviews of specific aspects of food science and technology with an emphasis on industrial relevance.

The *Key Topics in Food Science and Technology* series of short books therefore sets out to describe some fundamentals of food and food production and, in addressing a specific topic, each issue emphasises the principles and illustrates their application through industrial examples. Although aimed primarily at food industry recruits and trainees, the series will also be of interest to those interested in a career in the food industry, food science and technology students, food technology teachers, trainee enforcement officers and, established personnel within industry seeking a broad overview of particular topics.

Leighton Jones
Series Editor

PREFACE TO THIS VOLUME

All biological materials - including food - naturally deteriorate. Preservation technologies give us some control over this natural deterioration in food, allowing us to even-out our food supply over time. They enable us to use raw materials and products in line with need rather than production. Many food preservation technologies are well established and new approaches continue to emerge.

This book looks at the main approaches to food preservation that have become established on an industrial scale. It covers both traditional technologies, such as heat preservation and drying, as well as more recently established technologies such as modified atmosphere packaging. The aim is to describe the principles on which each technology rests and to illustrate these through examples and discussion of typical products in the marketplace. As with all the books in this series, the intention is to be illustrative rather than comprehensive. It also presents the information with a strong industrial slant to highlight how the fundamentals of food preservation are essential to securing the supply and distribution of safe and innovative food products.

Tim Hutton

ACKNOWLEDGEMENTS

Thanks are due to all CCFRA staff, past and present, whose reports I have used in the compilation of this overview, and specifically to Drs. Leighton Jones and Helen Brown for their comments on the text. Thanks also to Janette Stewart for typesetting the document, and especially for the design of the figures.

NOTE

All definitions, legislation and codes of practice mentioned in this publication are included for the purposes of illustration only and relate to UK practice unless otherwise stated. Specific products or companies are also mentioned at various points in this book, again for illustrative purposes, and this does not imply endorsement by CCFRA.

CONTENTS

1. INTRODUCTION

There is a limit to how long any particular food will remain fit for eating in general, or fit for a particular purpose. This is because of microbial changes (e.g. the growth of food poisoning or food spoilage organisms), chemical changes (e.g. the development of rancidity in fatty foods), or physical changes (e.g. water migration from a meat pie into the pastry crust, or the uptake of moisture by biscuits - in both cases making the product go soggy). In many cases microbial, chemical or physical changes will result in one or more sensory changes (e.g. off-odours or flavours, or texture and colour changes).

The limit to how long a product will last is termed its shelf-life. It is generally an advantage for a food product to have as long a shelf-life as possible: this means that it can be available for storage, distribution and consumption for a longer period of time. It is a continual goal of the food industry to increase the shelf-lives of existing products, without adversely affecting their quality, by modifying and developing new preservation steps. In some cases, the preservation step, which yields a product that has a relatively long shelf-life, is an intrinsic part of the overall processing of the food; i.e. the food does not exist without this processing/preservation step. For example, cheese and fermented milk products were developed as foods with intrinsically longer shelf lives than raw milk. Many food products, however, receive preservation processes that do not radically alter their nature or appearance; e.g. the shelf-life of milk can be extended to a certain extent by pasteurization, and to a much greater extent by sterilization, but the preserved product looks much like the raw product.

Food preservation is not a modern phenomenon. Mankind has always endeavoured to extend the shelf-life of foods, to try and ensure a year-round supply - balancing times of plenty (e.g. immediately after harvest) with times of shortage (e.g. during winter and early spring). Exactly what types of preservation were appropriate, and when they were needed depended on the lifestyle of the community. Even before farming and agricultural trading began in any formal way, when mankind was a hunter-gatherer, people were inventive about what they did with the food they gathered. Meat that could not be eaten after a kill was stored (either before or after cooking) by such methods as freezing, smoking or salting. (Freezing was carried out in the northern parts of Europe by burying the food in the permafrost). In hot climates, some raw materials could be left to the effects of natural microbial fermentation to produce a relatively stable, and safe, end-product. With the advent of farming communities, a year-round supply of animal products became less of an issue, but fruit, vegetable and grain products would still have been seasonal and a portion of these needed to be processed into a more stable form. Only in the last 30 years or so have we been able, through international transport and trade, to obtain fruit and vegetables in their raw state virtually all year round, rather than relying on preserved alternatives for part of the year.

Different cultures developed a variety of processed, preserved foods depending on the natural resources available to them, the amount of trade they could carry out, and the environmental conditions where they lived. This has resulted in the development, for example, of a wide variety of fermented, salted and cured meat and fish product, and an almost innumerable number of cheese and fermented dairy products. In 450 A.D., Attila the Hun reportedly preserved meat by storing it under his saddle, where sweat from his horse provided ideal preservation and processing conditions. History doesn't relate what it tasted like!

From the start of modern civilisation until the 19th Century, the long-term preservation of food relied on processes which significantly altered the characteristics of the raw material to yield a product that was inherently ambient-stable, and generally very different from the starting materials. This led to the development of various cured, salted, pickled, fermented and smoked products, as well as jams and other high-sugar products, and a variety of dried foods. Apart from jams and foods with very low amounts of water, the shelf lives of most of these products would be measured in weeks rather than years.

The shelf-life of the more perishable products could be extended to some extent by keeping the food in cool larders (i.e. by using lower temperatures to effect a degree of preservation). This short-term preservation method was in common use in the UK until the 1960s. Physical barriers such as fly screens were also used to keep out flies and similar pests to aid this preservation - a simple way of preventing contamination. It was in 1665 that Francesco Redi, an Italian physician, showed that the development of maggots in meat was not a spontaneous occurrence, but was in fact caused by flies being able to deposit their eggs in the meat. Protecting the meat with fine gauze prevented the entry of the flies and the growth of the larval stage - the maggot (Stanier *et al.*, 1976).

The first technology to be developed that could be applied across the whole spectrum of food types to give a long-shelf-life end-product was the canning of heat processed foods in an hermetically sealed container. In simple terms, if enough heat could be applied to a sealed container filled with food, all micro-organisms and intrinsic enzymic activity could be destroyed. The absence of microbial and enzymic activity coupled with a very low oxygen content meant that the three principal causes of food deterioration - microbial activity, and biochemical and oxygen-mediated chemical changes, could be eliminated. This technique could be applied to virtually any type of edible, raw product that could physically withstand the heating process. The technology of canning was initially developed by Nicholas Appert using corked, wide-mouth glass bottles and, in 1810, Peter Durand was granted a patent for the techniques using sealed metal containers (Lopez, 1987). Napoleon Bonaparte was an early purchaser of canned foods in metal containers for his troops. However, it was not until the 1920s that the canning industry became firmly established in the UK, when the need for long-term preservation of foods, principally for troops in battle, prompted the authorities to fund research into the available alternatives. Scientists from the Campden Research Station (then a Department of the University of Bristol - now CCFRA) visited the USA, where the canning industry was well established. They quickly realised that canning was likely to be the best way of preserving the large volumes of fruit and vegetables that were capable of being produced in the UK. (For further reading on the history of food preservation, see Thorne, 1986).

The advent of the refrigerator in the 1950s was the catalyst for the commercialisation of large-scale short-term preservation of products. Perishable food previously stored in cool larders could now be kept more conveniently, more safely and for an extended period. The growth in fridge ownership was very rapid -

in 1955, only 8% of UK households had them, but by the early 1960s, they were present in nearly all households. They usually had a small freezer box for storing ice-cream and/or frozen peas and burgers, but the variety of frozen food available was limited.

Although the technology of food freezing in the UK had been developed in the 1950s, it was not until the 1970s that the ownership of stand-alone freezers became widespread. This allowed the potential of food freezing as a means of long-term preservation of a wide variety of foods to be realized, and the huge array of frozen foods, including ready meals, to become available. Somewhat ironically, this was followed by the rapid development of chilled ready meals, in the race to provide more convenient food of improved quality.

Although these three techniques - canning, chilling and freezing - provide three of the main historical cornerstones of industrialised food preservation, they have not occurred in isolation. Pasteurization techniques, and chemical, microbiological and physical methods have all been developed, many from ancient knowledge, to result in the wide array of food types now available.

Methods of food preservation

The aim of food preservation is to prevent (or delay) the changes that occur in foods that make them unfit for their particular purpose. It is worth emphasising the causes of these deleterious changes:

- Micro-organism growth and associated activity (e.g. spoilage or toxin production)
- Enzyme and other metabolic activity inherent in the food
- Chemical changes caused by external factors (oxygen is the primary reactive agent)
- Physical changes caused by external factors (e.g. drying out of the food)
- Physical changes within the food (e.g. moisture migration from one component to another)

The type of preservation techniques that can be employed to prolong the shelf-life of a food will initially depend on the factors that limit its shelf-life. Thus, destroying indigenous micro-organisms with heat may be appropriate for products that are subject to rapid microbial spoilage, but would not be appropriate for preventing the development of oxidative rancidity in fatty foods.

The main preservation mechanisms can be summarised as:

- application of heat to destroy unwanted micro-organisms and enzyme activity
- application of cold to slow down chemical, microbiological and some physical reaction
- drying of food to prevent microbial growth
- addition of 'chemicals' or ingredients to slow down microbial growth, and physical and chemical changes
- addition of beneficial micro-organisms to compete with unwanted species
- creating an artificial atmosphere around the food, or removing the atmosphere altogether, to slow chemical and microbial changes
- provision of physical barriers in the food to prevent or slow down changes
- hurdle technologies combining two or more of these techniques

In addition to these major categories, other types of preservation techniques, such as the use of high pressure, have been developed. These are looked at briefly here, but are discussed in detail in Leadley *et al.* (2003).

The type of preservation mechanism that is used to prolong the shelf-life of the food may fundamentally alter the nature of that food. At the very least, it will subtly change its characteristics. This 'new' product will itself have factors that limit its shelf-life and these may well be different to those relevant to the 'original' food. For example, as shown in Table 1, the initial factor limiting the shelf-life of raw meat may be microbial growth; curing is one way of slowing this, but cured meats themselves may be subject to rancidity. This may be prevented or at least delayed by vacuum packaging. Depending on the nitrite, nitrate and salt levels in the cured product, the vacuum packaging may then require the product to be refrigerated to prevent the growth of the pathogen, *Clostridium botulinum*. At each stage, something will limit the overall shelf-life - nothing will last for ever.

**Table 1 - Example foods, shelf-life limitations and
potential solutions**

Food type	Potential shelf-life limitation	Possible solution
Fresh milk	Microbial spoilage	Pasteurization and chilling
Fruit and vegetables	Over-ripening, senescence and microbial rotting	Canning, drying, fermentation
Fresh-cut fruit and vegetables	Browning due to enzymic action	Modified atmosphere packaging
Meat	Microbial spoilage and pathogen growth	Curing, drying, canning, fermentation
Cured meat	Rancidity	Vacuum packaging
Nuts	Rancidity	Modified atmosphere packaging
Soft drinks	Microbial spoilage	Preservatives
Fruit juice	Microbial spoilage	Pasteurization and chilling, or sterilization
Paté	Rancidity	Vacuum packaging, antioxidants

Each of the preservation mechanisms utilised in food manufacturing needs to be carefully applied and monitored. Each has critical areas in which food safety or quality could be compromised if the process was not correctly controlled. A few examples are given in Table 2. These are not exhaustive, but are typical for the types of process mentioned.

In this Key Topic, the different preservation systems in widespread use will be summarised, with specific examples of their use.

Table 2 - Critical factors in food preservation mechanisms

Preservation technique	Some critical areas to monitor
Canning	• Correct amount heat applied • All parts of products in all cans receive adequate process • No contamination of can contents after processing
Pasteurization	• Correct amount of heat applied • Correct storage of product after process (e.g. refrigeration) • Adequate internal factors (e.g. salt and acidity) for intended shelf-life
Chilling	• No temperature abuse during production, distribution and storage
Freezing	• No temperature abuse during production, distribution and storage
Drying/Water activity control	• Low enough water content • Adequate salt/sugar etc. levels
Pickling	• Adequate acidity levels
Fermentation	• Correct starter cultures and conditions • Correct conditions during process to ensure stable end product
Modified atmosphere packaging	• Correct gas combination for food product • Correct permeability of packaging material • Adequate seals between packaging components
Vacuum packaging	• Adequate impermeability of packaging material • Adequate seals between packaging components

2. HEAT PROCESSING

The primary aim of heat processing is to preserve food by destroying micro-organisms inherent in the starting materials. Usually this is combined with some sort of sealing of the food from the outside environment to prevent recontamination immediately after processing. Heating also destroys enzymic activity in the food, thereby slowing or preventing some intrinsic biochemical deterioration. However, other, extrinsic chemical changes (e.g. oxidation or hydrolysis of fats) will largely be unaffected (some may even be enhanced), unless some other preservation mechanism is in operation (either as part of the heat processing preservation, or in addition to it).

Heat processing can simplistically be divided into sterilization processes and pasteurization processes. The former typically involves processes equivalent to 121°C for at least 3 minutes, and the latter temperatures of 70-100°C for a few seconds or a few minutes. Sterilization involves destruction of all organisms that could subsequently grow in the finished product, whereas pasteurization usually only completely destroys vegetative, pathogenic organisms. Also, sterilized foods can usually be stored at room temperatures for long periods of time (months or years), whereas pasteurized foods generally have shelf lives of days or weeks, when kept refrigerated (to prevent or slow the growth of spoilage organisms that have survived the heat process), unless some other preservative factor has been introduced. However, these categorisations are only generalities - some products can be kept at room temperature for long periods of time after having been given a pasteurization-type process. Other products will last for several months if kept chilled. The nature of the product, and the organisms that could survive and grow in it, need to be considered, as do other factors such as non-microbiological spoilage issues.

2.1 Sterilization processes

There are two basic ways in which thermally sterilized foods can be produced: heating the food after it has been sealed in a container (canning), or heating both the food and the container separately, and subsequently filling the food into the container under aseptic conditions (aseptic processing).

Canning

The principal concept of food canning is to heat a product in an hermetically sealed container so that it is commercially sterile at ambient temperatures; in other words, so that no microbial growth can occur in the food under normal storage conditions at ambient temperature until the package is opened (see Box 1). Once the package is opened, the effects of canning will be lost, the food will need to be regarded as perishable, and its 'shelf-life' will depend on the nature of the food itself - for example, the shelf-life of an open can of fruit may well be less than that of the raw fruit. Although the canning process traditionally employs metal cans, it is equally applicable to food sealed in glass and plastic containers, and, increasingly, in flexible pouches.

Box 1 - Commercial sterility

Commercial sterility is a very important concept in the food industry, specifically in heat-processed food. It differs from a medical definition of a sterile environment, in that micro-organisms may be present in the food. However, any that are present are incapable of growth under the combination of storage conditions and food matrix prevailing, so the food can be considered to be sterile.

In traditional canned food, the aim is to use a time/temperature combination for heating that will effectively eradicate spores of *Clostridium botulinum*. These combinations will not destroy some extremely heat-resistant spores of thermophilic spoilage organisms that may be present. However, these spores are dormant below about 50-55°C and therefore are not issue in most normal situations. Food processors would have to be wary, however, if the product was destined for very hot, sunny climates such as the Middle East, where direct sunlight on the cans might result in these temperatures being achieved.

Box 2 - Factors that affect the degree of heat process required

In addition to the many factors that affect the rate at which heat penetrates into a canned food product (see Box 4), there are several factors that affect the degree of heating that a product may need. These can be broadly summarised into two categories:

- The food product may offer a thermal protective effect to the micro-organism, thus necessitating a greater degree of heating (i.e. higher temperature or longer time of heating) to destroy all spores of *Clostridium botulinum*. It may also be highly contaminated with micro-organisms and similarly need a greater process (the destruction of microbial cells is essentially exponential, and therefore a greater heat process is required to reduce the number of spores to the stage where the chances of a spore surviving are 1 in 10^{12}).

- In contrast, the food may inhibit the growth of *Clostridium botulinum*, which cannot germinate and grow in highly acidic products, or if water activity is too low (water activity is a measure of the availability of water for biological systems, and is reduced by the presence of salts, sugars and other solutes). Also, preservatives such as nitrite (which is present, for example, in canned ham) will prevent the growth of the organism. For this reason, canned ham may only need to be given a pasteurization level heat process. Fruit products, being acidic in nature, can also safely be given a reduced process, which will help in maintaining the texture of the fruit. However, with unpitted stone fruits, there is an additional necessity - to destroy the emulsin enzyme system that could potentially give rise to the formation of cyanide precursors (see Box 3).

Clearly, it is very important to ensure that the correct heat process is given to any particular food. In a well-documented incident in 1990, there was a botulism outbreak associated with hazelnut yoghurt. The hazelnut puree used in the yoghurt was a canned product. The company involved in the canning of the hazelnut puree was experienced in the canning of fruit purees. However, when they introduced a hazelnut puree line, they failed to take enough account of the lower acidity of hazelnut puree, and as a result, the degree of heating applied was not sufficient to reliably destroy spores of *Clostridium botulinum*. In one particular batch at least, the spores survived and germinated, and subsequently produced the botulinum toxin. These were transferred to the yoghurt, which is not further processed before being eaten, and an outbreak of botulism resulted, including a number of deaths. The canning company involved went out of business as a result of the incident.

Box 3 - Cyanide formation in canned fruit

The cherry family, which includes plums, peaches, apricots and almonds, contain cyanogenic glycosides in their stones. One such glycoside is amygdalin. The stones also contain an emulsin enzyme system that can hydrolyse the glycoside and release cyanide. In the canning of whole stone fruit (i.e. with kernels remaining), there is the risk that enough of the emulsin enzyme system will remain, resulting in amygdalin hydrolysis and cyanide release and accumulation. The product therefore needs to be heated sufficiently to inactivate the enzyme. However, the situation is complicated by the fact that an excessive heat process will result in overprocessing and the fruit going mushy. The emulsin system in the stones of canned plums (termed beta-glucosidase, although it was subsequently found to consist of at least three enzymes working in sequence) was investigated by Haisman and Knight (1967). They found that the enzyme was thermally insulated to some extent. It was also stabilised by its isolation from the acidic constituents of the plum flesh, and by the relatively low water content and the high concentration of its substrate, amygdalin. After a canning process of 6 minutes at 100°C, cyanide content in the syrup rose to a maximum of 2ppm before declining; however, this level does not present a health hazard.

Hershkovitz and Kanner (1970) found that the enzyme system in the kernels of canned apricots remained active for long periods after processing if the heat treatment was not sufficient to inactivate it. A process of 86°C for 20 minutes (which was inadequate to destroy the enzyme) resulted in hydrogen cyanide levels in the cans increasing to 16ppm after 150 days' storage, after which it remained practically constant. A process of 38 minutes at 86°C resulted in levels of only 1ppm, which did not change throughout storage. As well as the potential toxicity of the raised cyanide levels, canned apricots containing more than 6ppm hydrogen cyanide were unacceptable from a sensory point of view, having too strong an almond flavour.

Nowadays in the UK, most stone fruit are pitted before canning.

Further reading:

Haisman, D.R. and Knight, D.J. (1967) Beta-glucosidase activity in canned plums. Journal of Food Technology 2: 241-248

Hershkovitz, E. and Kanner, J. (1970) The effect of heat treatment on beta-glucosidase activity in canned whole apricots. Journal of Food Technology 5: 197-201

The times and temperatures employed in sterilization operations will easily destroy all vegetative microbial cells. However, micro-organisms can develop structures called spores, which are resistant to heating, desiccation and other environmental stresses. The most heat-resistant pathogen spores that might survive the canning process are those of *Clostridium botulinum*. These may subsequently germinate in the absence of oxygen and produce a highly potent toxin. As the canning operation usually generates anaerobic conditions (i.e. no oxygen), all canning processes take into account the need to destroy these spores. In practical terms, the process applied must reduce the chances of a single spore surviving in a can to one in one million million (i.e. 1 in 10^{12}). This is called a "botulinum cook", and the minimum process for low-acid foods is equivalent to 3 minutes at 121.1°C. However, it is important to recognise that many factors inherent in the food may either allow a reduced level of processing to be applied, or necessitate an increased degree of processing (see Box 2 and Box 3). A process of F_0 5-6 is often applied.

Although *C. botulinum* is the benchmark organism for producing a safe product, there are more resistant spoilage organisms that can survive the process, such as *Bacillus stearothermophilus*. However, these tend to be thermophilic organisms, which will not grow under normal storage conditions, although they might become a problem if the food was to be kept in very warm climates. *B. stearothermophilus*, for example, will not grow below about 55°C. If the final product is likely to be subjected to this sort of temperature, it may be necessary to subject the food to a longer heating process.

In the traditional canning process, the product is filled into aluminium or steel cans, hermetically sealed lids are attached, and the cans are heated under pressure in a retort (an industrial scale pressure cooker). Heating is usually achieved by superheated water or steam, and care must be taken to ensure that the heat penetrates adequately to all parts of all cans and is distributed evenly, so that no 'pockets' of product are left underprocessed; the solids/liquid ratio and amount of 'headspace' are important in this. At the same time, it is desirable not to overcook the product, as this will result in reduced product quality. Therefore, much research has been undertaken to find the best ways of assuring heat penetration and of measuring it.

Box 4 - Heat penetration into canned foods

There are many factors that will affect the rate at which heat is transferred to canned food in a retort, and the time it takes for all parts of the food to be adequately processed. These include:

- the nature of the heating mechanism in the retort
- the number and arrangement of the cans in the retort
- the way that the cans are moved during the retorting operation - are individual cans agitated throughout the process, and are they inverted or otherwise moved during the process?
- the nature of the packaging - metal, glass or plastic; shape and size
- the nature of the food itself and how much of each component is present in the can - specifically, whether the food is likely to heat in the can by conduction (for solid foods), convection (for liquid foods) or a combination of the two (e.g. meat in gravy) - will the food 'move around' during the process, thus distributing the heat input more evenly? Also, how full is the can: air in a can acts as a significant thermal insulator
- the initial temperature of the food in the container

The aim of the operation is to ensure that an adequate level of processing is given to the food, while reducing, as far as is possible, any significant degree of over-cooking, which might be detrimental to product quality. The whole of the product formulation and processing criteria can affect the rate and uniformity of heat penetration, so it is important to calculate the likely kinetics for all new product formulations, and whenever there is a change in the nature of the packaging (e.g. from metal to glass), or in the dimensions of the container. Also, if the retort itself is changed, or its operating characteristics are altered, then heat penetration studies need to be carried out. Once the characteristics of the product and process have been calculated, and the desired regime is in place, the safety and quality of the end product is assured by accurate control and monitoring of the critical variables. End product testing can be kept to a minimum.

The determination of heat penetration itself is through a combination of mathematical modeling of the situation and actual measurements taken within different parts of the can at different locations within the retort. For temperature measurements, various types of device can be inserted into the can to determine the time-temperature profile. These measurements need to be taken from

continued....

wherever is calculated to be the cold spot in the can and the retort. Where this cold spot is located will vary, depending on product, process and retort characteristics: it is often the centre of the can in the centre of the retort, but there are many reasons why this may not be the case. Mathematical models can also be used to calculate the distribution of heat throughout the system, to try and avoid both over- and underprocessing.

Further reading:

There have been many generic studies published on the many factors that impinge on heat penetration in canning operations, and how to measure them. Among those published by CCFRA are:

May, N.S. (1997) Guidelines for establishing heat distribution in batch overpressure retort systems. Guideline No. 17

Smout, C. and May, N.S. (1997) Guidelines for performing heat penetration trials for establishing thermal processes in batch retort systems. Guideline No. 16

May, N.S. (2002) Analysis of temperature distribution and heat penetration data for in-container sterilization processes. CCFRA Review No. 22

Also of importance is the Department of Health publication "Guidelines for the safe production of heat preserved foods" (ISBN 011321801X)

After heating, the product needs to be cooled; it is vital that no post-process contamination occurs through the seal of the lid (see Box 5). Therefore, the integrity of the seal is vital and there are strict regimes for can handling. The water used to cool the cans must be of high quality microbiologically, and the cans must not be handled while wet as this could potentially result in contamination, with the water acting as a conduit for any micro-organisms present. The cause of an infamous typhoid food poisoning outbreak in 1964 in Aberdeen, linked with corned beef, was traced back to post-process contamination. The cans were cooled in the open air with water from the local river, downstream from a military fort where a typhoid outbreak had occurred! There are anecdotal reports of people still not buying corned beef because of this incident.

Box 5 - Can handling - avoiding post-process contamination

Canning is a highly successful way of producing long-shelf-life products from a wide range of ingredients. Its safety record is outstanding - there have only been a handful of food poisoning incidents in the UK in over 80 years. This level of safety has been achieved by well controlled processing operations, combined with appropriate handling of the cans after processing.

The nature of the metal can seal means that there is always the possibility that some will leak while the can is still hot, i.e. before the cooling metal contracts to form a completely tight seal. Ingress of micro-organisms is always possible at this stage if they come into contact with the can (resulting in 'leaker infection'). The two main routes of this are the cooling water and cannery personnel, but both are straightforward to avoid: the cooling water must be of high quality microbiologically, and the cans must not be handled whilst hot or wet. Once the cans are dry, the chances of contamination through the seam by manual handling is virtually eliminated - any micro-organisms on the can have no medium through which they are able to travel to enter the can.

These two basic rules need to be backed up by having well designed equipment to manoeuvre the cans after processing. The equipment needs to be easy to clean, and be kept clean, as transfer of microbial contamination from equipment surfaces, either directly or indirectly, needs to be avoided. The equipment also needs to be designed so that it is able to handle the cans without damaging them, as this could result in compromising of the can seam.

References:

Thorpe, R.H. and Barker, P.M. (1985) Hygienic design of post process can handling equipment. CCFRA Technical Manual No. 8

Thorpe, R.H. and Everton, J.R. (1968) Post-process sanitation in canneries. CCFRA Technical Manual No. 1

It is important that the cans are not damaged or deformed, as this could result in either perforation of the body of the can or damage to the lid seal. Either might allow the entry of micro-organisms into the can, thus negating the effects of the sterilization process, and resulting in the growth of post-process spoilage or pathogenic organisms. It is also possible that the lacquer applied as a coating to the inside of the can (e.g. to prevent the can from being attacked by acidic food components) might become cracked or damaged; this could result in the exposed metal dissolving into the contents of the can, leading either to a food tainting/contamination problem, or a thinning of the can wall and subsequent formation of pinholes.

Although cans are the traditional form of packaging for this type of process, glass, flexible pouches and semi-rigid plastic containers can also be filled and processed in a similar way. The use of plastics in particular has become significantly more widespread in recent years, and glass (which was the original material used in the process) is still used, particularly for 'high-value' products where being able to see the product is a marketing advantage. The term canning is still used to describe the process - in effect, it refers to the heating of product in a sealed container, not to the metallic nature of the container.

As with cans, proper sealing of glass or plastic pouches is important for preventing post-process leaker infection. The nature of the seal will be very different to that used in cans, but the basic rule of not handling the packages until cool and dry still holds. Some introductory information on metal, glass and plastic container seals is given in Hutton (2003).

Aseptic (continuous) processing

As in canning, the aim of aseptic processing is to produce a commercially sterile product through heat processing. The major difference here is that the package and product are sterilized separately and then the package is filled under carefully controlled aseptic conditions. In contrast to canning, which is a batch process, product is sterilized in a continuous process in which it travels through a heating pipe before entering a cooling leg and being transferred to the package. The product needs to be able to flow through the pipe easily and so aseptic processing is only suitable for liquids or suspensions of small pieces of solids in liquids (e.g. soups, or diced meat or vegetables in gravy). For such a system, hygienically designed and easily cleanable processing equipment is essential. Many of the issues addressed in the previous section on canning also apply to aseptically processed and packaged foods:

- the final package must be effectively sealed and the product must not adversely interact with the package
- the correct heat process must be applied to ensure that spores of *Clostridium botulinum* are destroyed or rendered incapable of germinating - this includes consideration of the rate of heat transfer into the food, and the characteristics of the food itself;
- the final product must be handled correctly to prevent post-process contamination

However, there are certain aspects that are of particular concern in aseptic processing. The primary one is that, unlike in canning, the equipment will come into direct contact with the food during both the heat processing and product cooling stages. This means that correct design of equipment is paramount. The same pieces of pipework will carry food over a long period of time, and at times the cooling leg will contain product at an ideal temperature for microbial growth. Therefore, equipment must be designed so that it can be cleaned easily between product runs, and there must be no 'dead-legs' where either food or cleaning material can become trapped. The design of liquid handling equipment in general is summarised in Hutton (2001), and is described in detail by Timperley (1997). As well as the physical aspects of cleanability, the material from which the pipes are made must be durable over the lifetime of the equipment, and be resistant to cleaning materials used; in addition, they must not subsequently transfer material to the food. Stainless steel is the material that is usually preferred. Similar properties regarding cleanability and non-transfer of constituents are required of all associated valves and gaskets that will contact the food. Again these are dealt with in summary by Hutton (2001), and in detail by Timperley (1997).

At the same time that the food is being processed, the packaging material must be sterilized. This can be done by using either heat (wet or dry), chemicals (e.g. chlorine, ethanol, hydrogen peroxide or peracetic acid), or UV or ionising radiation - [see Rose (1986) for further details]. This allows for a greater range of packaging types to be available compared with in-pack sterilized foods.

The sterile product then has to be filled into the sterile package without post-process contamination. Microbiologically speaking, this is a high-risk operation, as there will be no treatment step subsequently to remove or destroy any organisms that gain entry. The aseptic filling system may receive commercially sterile product straight from the sterilization unit, or it may be held in an aseptic tank or transported in bulk from elsewhere. If the filling system is connected directly to the sterilization system, it is important that the rate of packaging/filling does not exceed the rate of product sterilization, to ensure that under-sterilized product is not produced. Conversely, slower filling than processing may slow down the process line, resulting in overprocessing of product, and a drop in product quality. As with the sterilization equipment, it is vital that the filling equipment is hygienically designed and that it is regularly and effectively cleaned. The filling heads need to be compatible with both the product and the package. In some cases the filling head will have a special adaptor with which it positively engages the container to be filled. The filler needs to be able to dispense the correct amount of product into the package without any overspill onto the package seal area, as this would impede sealing and increase the risk of post-process contamination (Rose, 1986).

As well as the process/product opportunities with aseptic processing, there is also scope for the use of a range of different types of packaging, such as multi-layer cartons, plastic pouches and glass jars [see Hutton (2003) for examples]. Whereas with the canning process, the packaging has to typically be able to withstand water/steam combinations of around 120°C for extended periods of time, aseptic packaging is sterilized separately from the food and so the sterilization mechanism can be chosen to suit the packaging material. This enables materials such as paper and board-based composites to be used, which are popular and very convenient for long-life orange juice cartons. The packaging seals can also be designed to be compatible with both the packaging and the product, without the complicating factor of their stability at high temperatures and humidities.

2.2 Pasteurization processes

This is a heating regime (generally below 105°C) that aims to eliminate all pathogens (apart from spore forming organisms), as well as some spoilage organisms, from a foodstuff. The aim of using such a process is to generate an

Box 6 - Optimising product quality
in aseptic systems

The technique of aseptic processing is particularly suitable for fluid foods such as soups, and for fruit juices and similar products. One potential advantage is that of enhanced product quality, as the problem of overcooking can be significantly reduced by more accurate control of the heating process. Unlike in canning, where cans in different parts of the retort may receive a different amount of heat processing (and so all have to be processed long enough so that the slowest heating can is adequately processed), all food product is heated in the same place - the heating pipe. In its simplest form, this should mean that no part of the product has to be overheated in order for the rest to be adequately processed. Thus, higher quality products can be achieved without compromising food safety. However, as might be expected, this is an oversimplification. Even in a relatively narrow heating pipe, not all the food passing through heats at the same rate: liquids will heat more quickly than solids suspended in the liquid matrix. As well as controlling the rate of heat input into the heating stage, the rate and nature of product flow need to be monitored. Not all product will pass through the heating pipe at the same rate: solids will move more slowly than the liquids in which they are suspended and the liquids themselves will pass through at different rates. It was originally assumed that the flow of the liquid processed in these systems was broadly laminar and that it had Newtonian flow characteristics. In essence this means that the product flows in layers, and that the fastest flowing layer has a residence time of half of the calculated average flow rate (i.e. it receives half the average amount of heating that total product receives). This would mean that, in order for this section to be adequately sterilized, the whole process would have to receive approximately twice the 'average' amount of heat required to achieve commercial sterility. In experiments at CCFRA in conjunction with the Chemical Engineering Department at Cambridge University, it was found that this was not the case (Tucker, 1995), and that the assumptions were excessively conservative, resulting in unnecessary over-processing of the food.

Increasing knowledge of flow rate of liquids and solids in these systems, as well as conduction and convection rates in the system, will help the ultimate goal of aseptic processing, which is to produce even higher quality long-life products without compromising product safety.

end-product that has a significantly extended shelf-life, but with less marked quality changes, compared with the unprocessed food than is often apparent with sterilized foods. The actual degree of heat process required for an effective pasteurization will vary depending on the nature of the food and the nature and number of micro-organisms present, as well as the desired shelf-life of the product and the way in which it is going to be packaged and stored.

Milk is the most widely consumed pasteurized food in the UK. Pasteurization of milk was first introduced into this country in the 1930s, when a treatment of 63°C for 30 minutes was used. At the time, the process was not without controversy, with suggestions that the process could be used to 'make bad milk good' and was a licence for poor production hygiene and poor quality product. Modern milk pasteurization uses a process equivalent to a minimum of 72°C for 15 seconds (Singleton and Sainsbury, 1978; Singleton, 1997). Recently, there was some concern that the degree of pasteurization being given to milk might not always be sufficient to destroy the pathogen *Mycobacterium avium paratuberculosis*, which has been tentatively linked with Crohn's disease, a type of inflammatory bowel disease. As a precaution, many companies introduced an increased pasteurization time/temperature combination to their products to increase the assurance that this pathogen was destroyed by the process.

Pasteurization is now used extensively in the production of many different types of food, notably ready meals. Food may be pasteurized in a sealed container (analogous to a canned food), or in a continuous process (analogous to an aseptic process). It is important to note that pasteurized foods are usually not commercially sterile and will usually rely on other preservative mechanisms to ensure their extended stability for the desired length of time. Chilled temperatures are most commonly used, but some products have a high enough salt or sugar content to render them stable at room temperature (e.g. pasteurized 'canned' ham).

The vast majority of pasteurized products have moderately short shelf lives, and rely on chilled storage to maintain this. The following sections on chilling and freezing will look at the way in which low temperature affects the shelf-life of food. This section will concentrate on the heating regimes used in pasteurization processes and the primary aim of these processes, although reference to chilling and other preservative factors is inevitable as, unlike a sterilization heat process, pasteurization alone has a very limited preservative effect.

Box 7 - Cravendale milk filtration system

Recently, Arla Foods introduced a pasteurized, filtered milk (Cravendale PurFiltre) which it markets with a significantly longer shelf-life than conventional pasteurized milk products. It claims that a 20-day shelf-life is achievable in unopened containers. It also claims that its product has an improved flavour and mouthfeel over other milks.

The filtration system reduces by ten-fold the number of spoilage micro-organisms remaining in the milk compared with conventional pasteurization alone. In this way, ultimate spoilage of the milk is delayed. The system has no effect on the nutrient content of the milk.

The product was originally introduced in 1998, before being trialled extensively in London in 2001, and subsequently being launched across the UK in June 2002. It is now available through most major retail chains.

Although the product has this extra processing/preservation step, it still relies principally on the combination of pasteurization and chilled storage for its shelf-life.

The primary aims of pasteurization are to destroy vegetative (i.e. non-spore-forming) pathogenic organisms, and reduce the numbers of spoilage organisms significantly, as well as to destroy enzyme activity in the food that would also cause spoilage of the product.

Microbial destruction

The destruction of microbial pathogens is the key aim of pasteurization processes. If numbers are not reduced to negligible levels, then food poisoning could result, even if conditions do not permit their growth. Reducing food spoilage organisms is also important, in that it enables the food to be kept for longer. However, it may be sufficient to reduce their numbers to the point where, in combination with other preservation factors (such as chilled storage), they cannot multiply quickly enough to spoil the food within the given shelf-life. It is important to note that the existence of spoilage organisms in an unspoiled food is not a problem *per se*: they are not harmful. The length of time for which a food needs to be processed to achieve the desired pasteurization will depend upon the temperature of processing, as well as on the nature of the micro-organism and the prevailing environmental conditions (e.g. nature of the food, including its fat content, acidity, water content etc.). Under a given set of conditions, the length of time at a specified temperature which will result in a ten-fold decrease in microorganism numbers is termed the D value, and the number of degrees that results in a ten-fold change in D value is termed the z value (see Box 8).

The generally accepted situation is that, except in certain circumstances, a temperature of 70°C for 2 minutes will reduce the levels of all the major vegetative pathogens (*Escherichia coli, Salmonella, Listeria, Campylobacter* etc.) by a factor of 10^6 (a 6-log reduction); i.e. from an initial population of 1,000,000 organisms, only one will survive. Pasteurization protocols are based around the need for this level of reduction in micro-organism numbers (in practice a process equivalent to 72°C, rather than 70°C, for 2 minutes is often used as an extra precaution). However, different strains and populations of different organisms or mixtures of species may behave very differently, and the food in which they are found may have a very significant effect. For example, fat may provide a protective barrier in some situations. Therefore, the degree of pasteurization given to a food needs to be carefully calculated. As with sterilization regimes, the overall objective is to reduce the degree of heat processing applied, in order to maintain product quality, without compromising food safety.

Not only does the D-value change with temperature, and with the environment in which the organism finds itself, so does the z-value (i.e. the rate of change of susceptibility to temperature is not linear). Thus, great care has to be taken to ensure that each particular pasteurization process matches the foods being preserved.

Enzyme inactivation

In addition to microbial destruction, pasteurization processes often serve to inactivate the intrinsic metabolic processes in food raw materials, primarily raw fruits and vegetables. Enzyme activity in foods can be destroyed in much the same way as microbial activity.

Box 8 - Calculating heat processing requirements

When calculating the degree of heat processing needed to reduce micro-organism numbers by the required amount, there are two fundamental properties of the micro-organisms that need to be taken into account.

The first is the 'D' value, which is the amount of time required at a given temperature to reduce the numbers of micro-organisms 10-fold (i.e. 90% or one log order) (see Figure A). Thus a D-value of 30s at 70°C means that microbial numbers will be reduced by 90% in 30 seconds, by 99% in one minute, and by 99.9% in 1½ minutes. This value will differ greatly from one organism to another, and under different environmental conditions.

Figure A Example of a typical survivor curve used to calculate the decimal reduction time (D value)

Figure B Example of a typical death curve used to calculate the z value

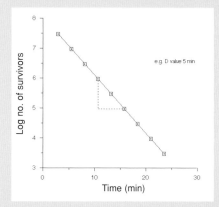

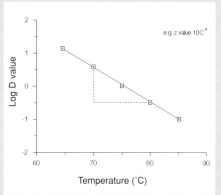

The second parameter that is extremely useful in calculating processing regimes is the 'z' value. The susceptibility of micro-organisms to heat is such that increasing the temperature by a few degrees will significantly reduce the length of time required to achieve the same reduction in microbial numbers. The z-value is a measure of this change and is the number of Celsius degrees required to increase the microbial destruction rate 10-fold (see Figure B). Thus, a z-value of 5C° means that increasing the processing temperature by 5°C will result in a 10-fold quicker destruction of micro-organisms; conversely, a 5°C reduction in temperature will

continued....

mean that 10 times longer will be required to effect the required reduction in microbial numbers. This change in micro-organism susceptibility can be used to great advantage in foods that are susceptible to quality loss during heating: a short, high-temperature process is usually better than a longer, lower-temperature one for maintaining product quality.

From the D and z values, the lethality of a process can be calculated in a given situation. In pasteurization processes, this is often quoted as a 'P' value, thus:

$$P_{70}^{10} = 20$$

meaning that the required reduction in microbial numbers (usually a 10^6 reduction) at 70°C will take 20 minutes. Increasing the temperature by 10°C (the z value) will result in a 10-fold reduction in the process time required (i.e. to 2 minutes).

In sterilization processes, a similar term, F_0, is used. This assumes a process at 121°C, and relates to the process received by the slowest heating part of the product. The 'botulinum cook', to reduce *Clostridium botulinum* spore numbers by 12 orders of magnitude, is regarded as 3 minutes, i.e. $F_0 = 3$.

Enzymes are proteins and rely on their 3-dimensional structure for their activity. This structure can be readily changed, and thus the activity destroyed by heating. Like micro-organisms, enzymes have different inactivation curves, and may be protected from the pasteurization regime by components of the food. Unlike micro-organisms, foods will not usually be 'reinfected' with a significant amount of enzyme activity after it has been destroyed, although it is possible for activity to be partially regained if the proteins have not been completely denatured. For most foods that are pasteurized, spoilage due to enzyme activity is not the shelf-life-limiting event: microbial spoilage is the main concern.

Examples of pasteurization

There are many types of food for which pasteurization is the primary preservation step. These range from chilled ready meals, through patés and sauces, to milk and fruit drinks and alcoholic beverages. Table 3 gives some examples, along with the critical micro-organisms for each product that the pasteurization step must destroy, and the additional preservative factors employed for stability. Psychrotrophic *Clostridium botulinum* is worthy of note. As has been discussed in the section covering sterilization, *C. botulinum* produces spores that are very heat resistant. Under anaerobic conditions (after processing) they can germinate and produce a lethal neurotoxin. In pasteurized foods, germination can be prevented by a number of factors, chilled temperatures being one. However, there are strains of *C. botulinum* that can grow at refrigeration temperatures (termed psychrotrophic). If there are no other factors to prevent their growth (such as low pH, or low water activity), these strains will grow and produce toxin. Fortunately, these strains are not as heat resistant as their mesophilic counterparts and can be destroyed by pasteurization regimes equivalent to 90°C for 10 minutes.

Like sterilization operations, it is essential that there is no contamination of the product with pathogens after the pasteurization process. There is no substantive difference between pasteurized and sterilized foods as regards post-process contamination issues. For products that are pasteurized in-pack, it must be ensured that the pack seals remain intact during and after the heating process and in subsequent handling. For those which are packaged after the pasteurization process, the same general rules can be applied as those for aseptically processed sterilized products: the packaging should take place in a high-care environment with suitably designed and maintained processing equipment that reduces the risk of contamination to negligible levels.

Box 9 - Interaction of pasteurization regimes with other factors

Mild heat treatments alone cannot usually assure the safety and stability of pasteurized foods. The treatment must, therefore, be used in combination with other factors. The most common of these are pH (acidity), water activity (including salt and sugar levels), undissociated organic acids (preservatives), modified atmosphere packaging and chilled temperatures. In most cases, however, there is the risk that

continued.....

these other factors may in some way be compromised, e.g. the food may be stored at slightly higher temperatures than are required for complete stability. Therefore, the heating process must be the major factor in conferring safety.

Depending on the natural properties of the processed food, and the subsequent storage regime, the pasteurization process given can be tailored to the individual needs of the product. For example, with products that have a pH of 3.8 or below, there is minimal risk of outgrowth of bacterial spores, even at ambient temperature. However, spoilage yeasts and moulds can grow under these circumstances, so the process will be designed to destroy these organisms. Above pH 3.8, acid-tolerant spores such as the butyric anaerobes can grow; these are more heat-resistant than the yeasts and moulds, so a greater pasteurization process will be required to ensure stability. Alternatively, combining low pH with low water activity will inhibit germination of this group of organisms.

However, there are factors that are outside the control of pasteurization processes. For example, the pasteurization process will not destroy spores of *Clostridium botulinum*, and so if the food is packed in a vacuum or in a modified atmosphere without oxygen, there is a risk that these spores will germinate and produce toxin. Low temperatures (i.e. refrigeration) will prevent or delay spore germination, but for products that rely on pasteurization and chilled temperature alone, a maximum shelf-life of 10 days is recommended. To achieve a greater shelf-life, other barriers to growth would need to be incorporated into the product: pH levels of below 4.5 and water activity levels below 0.94 generally prevent *C. botulinum* spore germination and growth.

In simple terms, the pasteurization process destroys existing vegetative pathogens and reduces the number of spoilage organisms; other factors are then relied upon to prevent the growth of resistant spoilage organisms and pathogenic spores.

Further reading:

Betts, G.D. and Gaze, J.E. (1992) Food pasteurization treatments. CCFRA Technical Manual No. 27.

Betts, G.D. (1996) A code of practice for the manufacture of vacuum and modified atmosphere packaged chilled foods with particular regard to the risks of botulism. CCFRA Guideline No. 11.

Table 3 - Examples of pasteurized foods

Food	Additional preservation factors	Significant micro-organisms for pasteurization step
Beer	Carbon dioxide	Yeasts
Brown sauce	Low pH (3.6); acetic acid salt; sugar; spices	Yeasts, moulds, lactic acid bacteria
Canned plums	Low pH (3.5); sugar (40%)	Yeasts; enzymes
Canned tomato sauce	Low pH (4.2); citric acid; salt; sugar	butyric anaerobes
Paté	Refrigeration; Low water activity (0.97)	*Listeria*
Milk	Refrigeration	Pathogens (particularly *Mycobacterium avium* subsp. *paratuberculosis*)
Meat products	Refrigeration; salt (2.5% in water phase)	Psychrotrophic *Clostridium botulinum*
Pickles	Low pH (3.3-3.8); acetic acid; salt; sugar; spices	Yeasts; enzymes
Soft drinks	Low pH (3.2-3.5)	Yeasts
Tomato ketchup	Low pH (3.8); acidity; sugar; salt	Yeasts (*Zygosaccharomyces bailii*)

Reference:

Betts, G.D. and Gaze, J.E. (1992) Food pasteurization treatments. CCFRA Technical Manual No. 27.

3. LOW−TEMPERATURE PRESERVATION TECHNIQUES

Whereas the primary preservation mechanism of the use of heat is to destroy micro-organisms in food that would otherwise limit its shelf-life, and similarly to destroy deleterious enzymic activity, low temperature acts by merely slowing down microbial growth and the changes caused by these micro-organisms and enzymes. It also slows down externally mediated reactions, such as oxidative rancidity. The Q_{10} concept suggests that, as a rule of thumb, chemical (including biochemical) reaction rates are halved for every 10°C drop in temperature. This is a broad approximation, especially in biological systems, where enzyme structure is important: at higher temperatures, the structure eventually starts to alter (eventually breaking down altogether), and biochemical rates will slow as a result.

Although the difference in temperature between chilled and frozen conditions may be only a few degrees, the effect on the preservative effect is highly significant. Biological systems need fluid water in order to operate; if this is denied (i.e. the liquid water turns to ice) then the systems effectively shut down, thus potentially allowing the food to be preserved for long periods of time. In chilled conditions, although some biological processes or systems may be slowed to the point where they are effectively stopped, others will continue (albeit slowly) and so shelf-life may still be quite limited. *Listeria monocytogenes*, for example, is well known for its ability to grow at very low temperatures, down to about -0.4°C.

3.1 Freezing

Freezing of food does not render it sterile. Whereas the freezing process may reduce micro-organism populations by up to 50%, this is not significant in the overall microbial quality of the food. Once conditions are favourable, most micro-organisms in food can grow, so that any lethal effects of freezing would be quickly counteracted. It is also important to note that food poisoning organisms

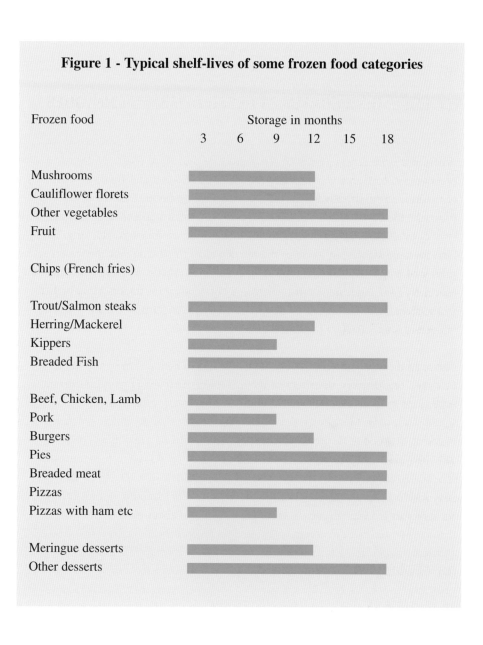

Figure 1 - Typical shelf-lives of some frozen food categories

or toxins present when the food was frozen will still be present when it is defrosted, and so a subsequent heat processing step (usually cooking) will be needed to ensure that the food is safe to consume. Frozen food that is not going to receive such a process (e.g. ice cream and frozen gateaux) needs to be of a suitable microbiological quality (i.e. free of pathogens) when it is frozen; thus, it or its ingredients probably need to have undergone a heat process or equivalent before the freezing process.

However, at the temperatures used in freezing processes (typically down to -18°C), all microbial activity is suspended and the length of time for which the food can subsequently be kept is dependent on other factors, particularly its texture and other sensory qualities such as rancidity. Nevertheless, frozen foods usually have shelf lives measured in several months to 2 years. Figure 1, which gives shelf-lives suggested by the British Frozen Foods Federation, demonstrates how effective the process of freezing is at preserving foods and extending their shelf-life. These are just guideline figures and individual circumstances may require a shorter shelf-life or allow a longer one.

The main factors that impinge on the quality and shelf-life of frozen food are the effects of the freezing process itself on the food, chemical and physical changes that occur during storage, and the thawing process.

The freezing process

The main issue that has to be dealt with in the freezing of foods is the effect that the formation of ice crystals has on the microstructure of the food. These will tend to disrupt cell integrity in unprocessed (i.e. raw) foods, which can be a significant problem, particularly for fruits. The rate of freezing is important as it directly affects the size of the ice crystals formed. Rapid freezing in blast freezers is desirable in order to prevent the formation of large ice crystals, which will increase any adverse effect on the texture of the product. Quick freezing is a standard industry process for many products, and many vegetable products especially are frozen using this technique to maintain as much of the original texture qualities as possible. This use of the term 'quick-frozen' is a legitimate marketing tool, but food labelled as such must comply with specific legislation (see Box 10).

Many fruits are unsuitable for freezing. As they often have a relatively delicate cellular structure, and a very high water content (usually over 90%), the formation of ice crystals can completely destroy the tissue structure, with subsequent defrosting of the product resulting in a pulpy mush, or a product which 'drips' water and dissolved sugars and colours. Vegetables tend to have a more fibrous structure, which is more resistant to the physical changes brought about by freezing, and individual quick frozen ('IQF') vegetables are a major food commodity.

Although the freezing process does have an effect on the physical properties of some foods, for many foods this is not a problem, and in the 1970s in the UK, the frozen food market saw widespread and rapid growth. Nowadays, a wide variety of high quality foods are available for retail sale either frozen at the point of sale, or chilled but marked as suitable for home freezing. Among the many food types that are regularly preserved by freezing are:

- ready meals
- pizzas
- raw meat and meat products
- bread
- vegetarian meat analogues
- sweet goods (e.g. cheesecakes, fruit pies)
- savoury specialities (e.g. Yorkshire puddings)
- vegetables

Chemical changes in frozen food

Even at freezing temperatures, enzymes remain in raw food such as fruits and vegetables, and could alter the sensory properties of the food. Although activity is very low in the frozen state, during freezing and thawing stages significant changes could occur. The potential problems with enzyme activity will depend on the food in question. For example, in peas, the enzymes responsible for lipid oxidation would, unless action was taken, quickly result in the development of rancid off-flavours; for this and related reasons, vegetable products are usually blanched before freezing to ensure that enzymes are denatured.

Box 10 - Quick frozen food labelling legislation

Because the quick freezing process can offer important benefits to the textural quality of many frozen foods, the use of the term 'quick frozen' in labelling or advertising of the food is controlled by legislation. The two key points regarding temperature are that the food freezing process must take the food through the zone of maximum crystallisation as quickly as possible, and that it must be cooled to minus 18°C or colder and be maintained there (although in transport and retail display the temperature may rise to minus 15 and minus 12°C respectively, consistent with good practice). The freezing must be carried with sufficient promptness and with appropriate quickness to minimise any chemical, biochemical or microbiological changes to the food.

In the freezing process, only air, nitrogen and carbon dioxide can be used as cryogenic media. The food must be clearly labelled with an instruction of the type "do not refreeze after defrosting". (As domestic freezers operate very slowly in comparison with commercial blast freezers, to do so would negate all of the quality aspects of the food maintained by the quick freezing process.)

These regulations only cover the labelling or advertising of food as 'quick frozen'; there is nothing to stop food processors from operating freezing processes differently - non-quick frozen food can be perfectly acceptable - the only restriction is that manufacturers cannot then use the term 'quick frozen' in connection with the food. In some cases, the potential improvement in quality from quick freezing may be slight and not warrant the extra costs associated with the quick freezing process.

Reference:

The Quick-frozen Foodstuffs Regulations 1990. SI No. 2615

Blanching is a pasteurization-type process designed to eliminate enzyme activity rather than microbial activity. A typical process involves heating to 90-95°C for a few minutes before rapid freezing.

Apart from enzymic activity, there are many other chemical changes which may limit the shelf-life of frozen food; one main example is non-enzymic fat oxidation, which may occur over a period of months, depending on the product. Although significantly slowed at freezing temperatures, fat oxidation and other chemical

Box 11 - 'Do not refreeze' labelling

Many frozen foods are labelled with the instruction to not refreeze after thawing. There are two main reasons for this: microbiological safety and textural quality. Foods such as raw meats naturally have a relatively high microbiological load. When such foods are frozen, the growth and multiplication of the organisms is stopped, but there is no significant decrease in microbial numbers. Once the product has been thawed, these organisms can start to grow again (albeit after a lag phase), so it is important that the item is cooked or otherwise prepared for consumption before microbial numbers proliferate to such an extent that the cooking process might be 'overstretched'. The danger with refreezing items is that, during the second freezing and thawing cycles (which in domestic situations can be fairly lengthy), there is an extra chance for microbial numbers to increase. As well as providing the extra risk of pathogen numbers increasing, spoilage organisms may also have an extra chance to detrimentally affect the sensory quality of the food. It must be stressed that this type of concern only applies to microbiologically high-risk foods, such as raw meats, and it is the length of time for which the product is thawed and the temperature that are of concern. To compare, although many fresh meat joints are labelled as suitable for home freezing, they usually have an instruction along the lines of 'freeze on day of purchase', and are given a relatively short raw shelf-life.

Of much wider applicability is the textural quality that may be lost by home refreezing. Commercial freezing processes are usually very rapid, causing minimal damage to the texture of the food. However, freezing in domestic refrigerators is a much slower process, and the formation of large ice crystals may adversely affect texture, either through cell structure damage in fruits and vegetables, or the destruction of emulsions and processed textures, such as in dairy desserts.

changes leading to rancidity development in fatty foods will continue during storage; as the fat is in the non-aqueous part of the food, it will be less affected by the phase change of water to ice. Other changes that can occur include fading or changes in the colour of the food, and changes in the flavour components. In fruits and vegetables, these can be triggered by freezing-induced damage in the tissues allowing the mixing of components that would otherwise be kept separate.

Physical changes in frozen food

The changes in physical texture that occur within frozen food are often the factors that limit the overall shelf-life of the product, especially in primary products such as meat, fish and vegetables. In domestic storage situations especially, even with well controlled freezers, there will be a significant amount of temperature cycling, with localized rises and falls in temperature within the food. This will have the effect of creating larger ice crystals within the product and so adversely affecting product texture - for example, by bursting cells or sub-cellular compartments, allowing reactions between components that would have been held separately, and increasing loss of liquid from the product on thawing (drip loss). This can have major effects on the eating quality of the food, and so limit the period for which quality is maintained.

Surface drying of the food will also occur over prolonged storage, and this may also have a significant effect on the texture and appearance of the product. It is likely, especially in domestic situations, that the surface of the food will regularly be in contact with moist air above 0°C. Two things are likely to occur as a result: slow evaporation of moisture from the food, and condensation and freezing of moisture from the air onto the food, resulting in a dry, frosty surface to the food ('freezer burn').

3.2 Chilling

Keeping products at a low temperature retards both microbiological and chemical deterioration of the food. In most processed chilled foods, it is microbial growth that limits the shelf-life; even the slow growth rates that occur under chilled conditions will eventually result in microbial levels that can affect the product or present a potential hazard. This microbial growth can result in spoilage of the product (it may go putrid or cloudy or show the effects of fermentation), but it is also possible for pathogens to grow with no noticeable signs of change in the food. It is important to note that food showing microbial spoilage is not necessarily poisonous. Conversely, just because a food has not visibly spoilt does not mean that it is safe, and it is important to pay attention to the use-by date on chilled foods.

To reduce microbial effects to a minimum, chilled prepared foods are usually given a pasteurization heat process, sufficient to eliminate pathogens such as *Salmonella, Listeria* and *Escherichia coli* O157. A process equivalent to 70°C for 2 minutes is generally considered to be sufficient, but the exact process given will depend on the nature of the food. For more information on this type of process, see the section on pasteurization [Section 2.2].

Chilling is also used to prolong the shelf-life of fresh fruit and vegetables. Here, low temperatures not only retard the growth of naturally occurring micro-organisms (which might rot the product), but also slow down biochemical processes (e.g. those associated with ripening and senescence) that continue after the product has been harvested. However, each individual fruit and vegetable has its own ideal storage temperature (see Table 4), and some are susceptible to chill injury. Bananas, for example, can not be stored chilled: at low temperatures, tissue damage occurs, releasing phenols and the enzyme polyphenol oxidase; in the presence of oxygen, this results in a very rapid blackening of the banana tissues.

Table 4 - Ideal chilled storage temperatures for fruit and vegetables

Temperature (°C)	Products
1-2	kiwi fruit
3-5	apples, apricots, avocado, cabbage, carrots, cauliflower, grapes, leeks, lettuce, nectarines, onions, parsnips, peas, pears, plums, raspberries, rhubarb, strawberries
0-8	clementines, mandarins, oranges, satsumas, tangerines
4-8	cherries, melons, squash
6-9	celery, cucumbers, pineapples, potatoes, tomatoes (ripe)
4-10	aubergines, courgettes, green beans, marrow, okra, peppers
10-13	ginger, grapefruit, mango, pumpkin
11-14	lemons
10-15	tomatoes (green)
13-18	bananas

Reference: Frozen and Chilled Foods magazine poster

Chilling is an excellent preservation mechanism for achieving a moderate extension of shelf-life above that which could be achieved by keeping the food at room temperature. As it does not involve the formation of ice crystals, the problems faced with loss of textural quality in the freezing process are avoided. However, unlike the freezing process, which suspends microbial activity completely and drastically slows chemical and biochemical changes, chilling is merely a process for moderately slowing down microbial, chemical and biochemical deterioration of foods. Its effectiveness will depend significantly on what other treatments the food has undergone, and the nature of the food itself. If there has been a heat treatment to destroy vegetative pathogens and the food is contained within a sealed package, then the length of time the food can be stored will depend on how long it takes for significant growth of remaining spoilage organisms to occur or for chemical or physical changes to cause the quality to deteriorate. (This assumes that overall conditions do not allow the germination and outgrowth of any remaining pathogenic spores, e.g. of psychrotrophic *Clostridium botulinum*.)

Sealed paté products, for example, often have a shelf-life of around 3 months; however, in common with many other packaged, chilled foods, they are labelled with the instruction along the lines of 'consume within 3 days of opening'. This instruction derives from two main concerns: the opening of the product to microbial contamination from the atmosphere; and the exposure of the product to oxygen. Fatty foods are very susceptible to the development of rancidity and other oxidative changes. In a closed package, oxygen levels are limited, but once the package is opened, there is a plentiful supply of the gas, which can rapidly promote the development of off-flavours, as well as colour changes, in the product.

Chilling effect on microbial growth

Most commonly encountered food microbial pathogens and spoilage organisms grow best somewhere in the range 20-40°C. Below about 8°C, growth of pathogens is significantly curtailed, and below about 3°C, in most cases it ceases entirely (although a few - most notably *Listeria monocytogenes* and *Yersinia enterocolitica* - can still grow at this temperature and as low as -1°C - see Table 5). Thus, in some situations, chilling below 3°C can be used to extend the microbial shelf-life of foods very significantly - the limiting factors may be external contamination with organisms that are capable of growth at temperatures between 0 and 3°C, or non-microbial changes,

Table 5 - Minimum growth temperatures for selected pathogens

Organism	Minimum Growth Temperature (°C)
Escherichia coli O157 (and other VTEC)	7.0
Staphylococcus aureus	6.7
Vibrio parahaemolyticus	5.0
Bacillus cereus	4.0
Salmonella species	4.0
Clostridium botulinum (psychrotrophic)	3.3
Yersinia enterocolitica	-1.0
Listeria monocytogenes	-0.4

Reference:
Betts, G.D. (1996) A code of practice for the manufacture of vacuum and modified atmosphere packaged chilled foods with particular regard to the risks of botulism. CCFRA Guideline No. 11.

such as drying out of the product, or chemical or texture changes. This fact is put to good effect in the preservation of some vacuum packed foods (see Section 8) - where vegetative pathogens such as *Listeria* and *Yersinia* have been eliminated by a cooking process and the product is packed in a hermetically sealed container. However, storage below 3°C can only be reliably maintained in commercial storage situations, and not during commercial distribution, retail display and domestic storage.

Chilling effect on rancidity development

The development of rancid off-flavours is a major factor in limiting the shelf-life of foods, especially those with a high fat content (see Box 19 - *Rancidity of Fats* - for a description of the main types of rancidity). The rate of rancidity development is highly dependent on the nature and composition of the food (e.g. whether it has antioxidants naturally present or added, and whether there are components present which promote rancidity) and the way it is processed (e.g. whether oxygen is incorporated, as happens in butter churning). However, temperature is also highly significant. Generally, lowering the temperature will significantly reduce the rate of rancidity development. Interestingly, freezing can result in an increase in the rate due to the concentration of solutes.

Chilling effect on fruit and vegetable metabolism

Fruits and vegetables, even after they are picked/harvested, continue to metabolise - as well as taking up oxygen and giving off water and carbon dioxide, they will continue to mature (ripen) and will eventually start to decay. Even if all outside agents of spoilage are inhibited, biochemical reactions internally will eventually lead to the product becoming unfit for consumption. Chilling can slow down these changes - initially allowing longer storage periods before the product reaches its ideal condition, and then keeping it near this peak condition for longer. This can be combined with modified or controlled atmospheres in packaged products to further enhance shelf-life. Some products have a very narrow temperature range for ideal storage and extension of shelf-life, whereas with others a combination of factors means that there is a relatively wide range of ideal temperatures (see Table 4, p. 35).

Box 12 - Monitoring temperatures in chilled storage areas

Because the shelf-life of chilled food is critically dependent on the temperatures to which the product is subject during manufacturing, transportation, retailing and catering, it is necessary to have a way of monitoring the temperature of food within such environments. Whilst it is relatively easy to monitor the temperature of a storage area itself, using conventional thermometers, changes in storeroom temperature will not necessarily be mirrored by immediate changes in product temperature. Until recently, this needed to be determined using temperature probes. However, modern thermal simulation systems (either an instrument or a computer system) can be used to mimic temperature responses of one or more foods within, for example, chilled food display cabinets, which greatly reduces the need for the destructive testing of foods using temperature probes. The system may consist of a 'dummy' food with temperature probes in suitable locations, or probes in the storage area itself linked to a computer programme that will predict product temperature within the store.

The simulation systems take account of the effects of size, shape and packaging on the cooling rates of products, and can be used for the continuous monitoring of the effects of localised temperature changes. The systems are cheaper than direct temperature probing systems, and they are very reliable: once properly calibrated, they will respond accurately and consistently to the same set of conditions. They also avoid problems with cross-contamination: the centre temperature in a piece of food can be mimicked without the need to probe real food.

Reference:

Tucker, G. (1995) Guidelines for the use of thermal simulation systems in the chilled food industry. CCFRA Guideline No. 1.

Box 13 - Achieving reasonable working conditions in chilled food areas

The production of chilled food is associated with a unique problem when it comes to worker health and safety. Because of their nature, chilled food ingredients usually need to be kept chilled during the production of the finished product. Food safety legislation requires chilled products to be kept below 8°C. However, workplace legislation and an associated code of practice stipulate a minimum temperature of 16°C for worker comfort.

Pragmatic measures need to be taken to try and satisfy, as far as is possible, these opposing requirements. This is done through a hierarchy of control involving engineering and hardware options, factory design and thermal protective clothing. The various 'engineering' solutions include enclosing or insulating the product, pre-chilling the product, keeping chilled areas as small as possible, and exposing the product to workroom temperatures for as little time as possible.

In addition to these engineering solutions, there are many issues to consider at the factory design stage; these include not situating workers in draughts or next to chill room doors, carefully designing the product flow from raw material to despatch, and analysing personnel access, movement and changing facilities. Consideration of air movements and manipulating them so that cold air is directed over the product line and away from the operatives is also very important. This has to be done in a hygienic way; air can be a major vector for contamination of food and food processing equipment and there are many precautions that need to be taken to ensure that it is of suitable quality (see Hutton, 2001; Brown, 2000).

The third (and 'last resort') option for keeping product cool and operatives warm is the provision of protective clothing. Exactly what clothing is required will depend on the specific circumstances, particularly the level of physical activity of the workers and to what degree and for how long they are subjected to sub-optimal temperatures. A code of practice (BS 7915:1998) looks at these issues.

Further reading:

- Food Safety (Temperature Control) Regulations 1995. SI 2200
- The Workplace (Health, Safety and Welfare) Regulations 1992. SI 3004
- British Standard 7915 (1998) Ergonomics of the thermal environment. Guide to design and evaluation of working practices for cold indoor environments
- Brown, K.L. (2000) Guidance on achieving reasonable working temperatures and conditions during production of chilled foods. CCFRA Guideline No. 26
- Hutton, T. (2001) Introduction to hygiene in food processing. CCFRA Key Topics in Food Science and Technology No. 4

4. DRYING AND WATER ACTIVITY

Micro-organisms need water to grow. Reducing the amount of water in a product that is available for the micro-organism to use is one way of slowing or preventing growth. There are two main approaches to this: physically drying the product, or preventing the micro-organism accessing the water that is present in a wet product. Thus, dried herbs and spices may be stable for many years. Many staple foods are available in dried forms (e.g. breakfast cereals, pasta and rice) and these will also remain fit for use for a long period of time. The shelf-life of breakfast cereals is usually limited by texture changes, with the product losing its crispness and becoming soft or 'cardboardy'.

The production of genuinely dry foods varies significantly with food type. It is outside the scope of this book to look at the whole production processes for these products, but brief details of the drying process itself are given for breakfast cereals in Box 15.

Some products may contain fairly high levels of water, but with much of it 'tied up' by the soluble constituents of the food (i.e. the product has a low water activity - see Box 14). For example, the sugar present in the jam reduces the water activity of the product, and the water present is not available for the micro-organism to use. As a result, traditional jams can be kept for many months without spoiling. Conversely, many low-sugar jams have to be refrigerated after opening, as the sugar levels are not sufficient to prevent microbial growth. This is a good example of how the consumer-driven desire for products with altered characteristics (i.e. jam with less sugar and no 'artificial' preservative) has resulted in a product that has lost one of its major, original characteristics - long-term stability at room temperature (see Box 16).

The other major ingredient that is used to lower water activity is salt. Salt can be used to preserve meat, fish, fruit and vegetable products, either as the sole preservative or in combination with acid, fermentation or both - salting of meat and fish in particular is a very old method of preserving foods. Salt can be added either in the form of dry salt, rubbed onto the surface of the food, or as a brine solution. Saturated brine is usually required to achieve long-term preservation if this is to be the only preservative factor. When the product is going to be consumed or further processed, the excess salt will usually need to be removed to yield a palatable product.

Reduction of water activity is a very significant mechanism for preserving food. As well as in the traditional preserved foods mentioned above, it is also a major factor in the development of many new products, such as sauces, and in the stability of products that are mainly preserved by other mechanisms. This combination of different preservation mechanisms is known as hurdle technology and is dealt with in more detail in Chapter 10. As an example, canned ham, which is a commercially sterile product, receives a heat process that is more equivalent to a pasteurization

Box 14 - What is water activity?

Water activity (a_w) or equilibrium relative humidity (ERH) is a measure of the availability of the water in a system to assist in the metabolic processes of organisms living in that system (Ranken *et al*, 1997). (ERH is water activity expressed as a percentage so: ERH [%] = a_w x 100.) The water activity-lowering effect of a solute depends, amongst other things, on the number of particles of solute in the solution: sugars, being much larger molecules than salt, have a lesser effect on a weight-for-weight basis. In theoretical situations, Raoult's Law applies:

$$a_w = Nw/(Nw + Ns)$$

where Nw and Ns are the number of water molecules and solute molecules respectively. As systems become more complex, containing many solutes, there are interactions between the solutes, and the above equation becomes an approximation. There are many types of hygrometer for measuring water activity, examples of which are given in Voysey (1999).

The water activity of pure water is 1.0; that of a saturated (68%) sugar solution is 0.85, and that of a saturated (26%) salt solution is 0.8. Many food poisoning bacteria are incapable of growth below an a_w of 0.95, and other bacteria are inhibited at 0.9. Examples are given in Table 6. Yeasts can grow at values as low as 0.8, whilst moulds can grow down to values around 0.7.

Table 6 - Examples of water activity thresholds and sodium chloride concentrations for growth of some micro-organisms (assuming all other conditions are optimal)

Micro-organism	Water activity min	Salt concentration max (%)
Acinetobacter	0.990	1
Clostridium botulinum Type E	0.970	4
Pseudomonas fluorescens	0.957	7
Clostridium perfringens	0.950	7
Escherichia coli (including O157)	0.950	7
Salmonella	0.950	7
Vibrio parahaemolyticus	0.945	8
Clostridium botulinum A and B	0.930	10
Bacillus cereus	0.910	13
Bacillus subtilis	0.900	15
Staphylococcus aureus	0.860	20

process. *Clostridium botulinum* spores, if present, will not be destroyed, but will be incapable of germination, growth and toxin production because of the salt, nitrate and nitrite levels present. Reduction in water activity is one of the principal mediators of their effect. Mayonnaises and sauces may be ambient stable or require refrigeration for their preservation, and ambient stability may be lost after opening. Water activity, along with acidity and other factors, is important in determining on which side of the dividing line a particular product falls.

In contrast to products with a high water content but a low water activity, other products may have a fairly low water content, but that which is present is freely available for microbial growth (e.g. cheese and bread), i.e. the product has a high water activity. If this available water is at the surface of the product, then spoilage from environmental micro-organisms will eventually occur. Although cheeses, for example, are generally less than 50% water, the salt level in the water phase is only about 2%, which is not high enough to prevent microbial growth.

Two of the main traditional chemical methods of preserving foods - smoking and curing - both rely on a reduction in water activity as well as a direct chemical effect for their preservative action. The specific characteristics of these processes are dealt with in Section 5, but are mentioned here in the context of water activity. The changes in cellular mechanisms in micro-organisms that result from changes in water activity are often closely linked to those that arise from 'direct' chemical preservation methods - it is often difficult to distinguish one effect from the other.

Box 15 - The drying of breakfast cereals

There are many different types of ready-to-eat breakfast cereals, made primarily from corn (maize), wheat, oats and rice. Fast and Coldwell (1990) describe the many types of processes involved in their production. Drying is the controlled removal of water from the cooked grain and other ingredients to obtain an intermediate with appropriate properties for further processing. In the same way that fermentation reactions in products such as cheese and yogurts are both an intrinsic part of the production of the food and the main method for its preservation, so the drying steps in cereal manufacture are essential in the production of the final product, as well as being the main method of preservation. In other words, the drying is an integral part of cereal manufacture, not merely something that is done at the end of production to preserve the final product.

The drying step may occur at several intermediate points in the production process. The cooked cereal mass may be predried to prevent agglomeration and subsequent damage during handling. The drying process is likely to result in a moisture gradient within the product as it leaves the drier - this needs to be addressed so that a more uniform moisture content is achieved within and among the cereal particles. Subsequent operations, such as puffing and toasting, involve further drying of the product, as well as changing its physical structure and chemical make-up. This may be followed by another drying stage after the application of, for example, a sugar syrup coating; at this stage the cereals are dried to set and harden the coating and remove excess moisture.

Reference:

Fast, R.B. and Caldwell, E.F. (1990) Breakfast cereals and how they are made. American Association of Cereal Chemists

Box 16 - Jam preservation

The fruit present in jam contains a significant amount of sugar, typically 6-10% depending on the particular fruit. However, for an ambient-stable preserved product, sugar levels of about 65-70% would be needed at the end of the process. This extra sugar is usually added mainly in the form of cane or beet sugar (sucrose). In jam manufacture, the raw fruit, sugar and pectin are boiled together. As well as reducing the total water level in the mixture, and causing texture changes in the fruit and pectin, this also results in some conversion of the sucrose to a mixture of glucose (dextrose) and fructose; this mixture of monosaccharides is known as invert sugar). This plays a small but significant role in the preservation mechanism - solutions of sucrose become saturated at about 66%, whereas mixtures of sucrose and 20-35% invert sugars only become saturated at 72%. Depending on the actual process to be employed, invert sugar or glucose syrup can be included in the starting recipe.

The boiling process will kill all vegetative pathogens and spoilage organisms and, at the prevailing sugar levels, all bacterial growth will be inhibited - yeasts and moulds are the micro-organisms of concern in the surface spoilage of jams. If sugar levels are lowered too much, then refrigeration may be required for prolonged storage after opening. The UK 'Jam and Similar Products Regulations 2003' state that 'normal' jams, jellies and marmalades must have a total soluble solids (sugars) concentration of at least 60%; lemon curd and similar products must have a total soluble solids content of at least 65%. Products containing between 25 and 50% total soluble solids must be labelled as 'reduced sugar' products, and are permitted to contain specified preservatives, required to replace the preserving effect lost with the sugar.

The other specific factor in the preservation of jams and marmalade is their physical and textural preservation - the prevention of crystallisation of sugars and avoiding syneresis in the product (i.e. the separation of solids and liquids). Crystallisation is avoided by attaining the correct levels and balance of sucrose and invert sugars and controlling the degree of boiling and subsequent water loss during processing.

The gelled characteristic of jam is derived from the formation of sugar-pectin gels. Pectin levels in raw fruit vary greatly, and so different amounts of pectin have to be added to different jams to effect the correct gelling consistency. The type of pectin added is also important and is affected by the processing method and the other constituents of the formulation (e.g. sugar levels, and presence of calcium ions and fruit acids) (Chapman and Baek, 2002). The right combination results in the familiar gelled texture of jam that is maintained throughout its shelf-life, even after repeated re-use of the product.

5. CHEMICAL PRESERVATION METHODS

The addition of specific chemicals to foods to inhibit microbial growth and chemical transformations is a major method of preserving food. It is not a new innovation - smoking, pickling and curing, for example, are traditional forms of preserving food that rely on the addition of a mixture of chemicals, primarily to inhibit microbial growth. This practice has led to a range of products with unique flavour and taste characteristics that rely largely on the preservation mechanism.

The chemicals added to food can either act directly on the food system, or indirectly via the water content of the food (i.e. by lowering water activity). Traditional systems such as smoking and curing rely on a mixture of these two effects. The effect that water activity has on food systems is dealt with in more detail in Section 4, p40. The addition of chemicals to food is carefully controlled by national and international regulations.

5.1 Preservatives, antioxidants and other additives

Of the classes of chemicals permitted to be added to foods, antimicrobial additives ('preservatives') probably receive most attention; there are relatively few of these permitted for use in the UK and EU and in many cases there are specific limits on how much can be used and in which products (see Table 7). The use of some preservatives is limited to just a few types of food (e.g. nitrate and nitrite salts to specific meat, cheese and fish products).

The two preservatives that are used most widely are sorbic acid and its salts and sulphur dioxide and its derivatives. There is currently a major consumer-led move to reduce the range of foods containing preservatives and the levels of preservatives actually used. This poses a significant problem for the food industry, as a reduction in preservative level either necessitates use of another preservation technique

Table 7 - Examples of antimicrobial preservatives permitted in the EU

Preservative	Typical permitted foods	Maximum levels allowed
Sorbate	Wide variety, e.g. drinks, jam, cheese, bread	Variable, but typically 1000-2000 mg/kg
Benzoate	Wide variety, but fewer than sorbate	Variable, but typically up to 500 mg/kg
Parahydroxybenzoate	Meat jelly coatings, savoury snacks and confectionery	Usually 300 mg/kg
Sulphur dioxide, sulphites, bisulphites and metabisulphites	Very wide variety of foods, e.g. burgers, sausages, dried fruit and vegetables, drinks	15-2000 mg/kg, depending on the food
Biphenyl	Citrus fruit surface	70 mg/kg
Thiabendazole	Citrus and banana surfaces	6 and 3 mg/kg respectively
Nisin/ Natamycin	Clotted cream, cheeses and semolina	3-12.5 mg/kg
Boric acid	Caviar	4 g/kg
Nitrites	Cured meats	150 mg/kg
Nitrates	Cured meats, cheese and pickled fish	300 mg/kg
Propionic acid and its salts	Bread products	1000-3000 mg/kg

Reference: The UK Miscellaneous Food Additives Regulations 1995. SI 3187.

(e.g. heating, chilling or freezing) or a significant reduction in shelf-life. Both of these alternatives may result in a product of actual or perceived poorer quality.

In deciding what preservatives should be allowed in foods, and in what foods and at what levels, several factors are taken into consideration by legislators in the UK and EU:

- the technical requirement (i.e. the need for a chemical preservative in that food)
- the economic requirement (that there is a real demand for the food in question)
- the availability of alternatives (and their efficacy and cost)
- the likely levels of consumption of the additive and its potential adverse health effects

In order for a novel preservative to become permitted, or for a new application of an existing preservative to be allowed, it is usually up to the company or companies potentially producing the food in question to justify the need. This is then evaluated by various expert committees, who also take into account the toxicology aspects. Preservatives are only permitted at levels which are required to have the desired technological effect, and which will not result in adverse health effects at levels likely to be consumed.

Food companies are then often faced with a decision on which preservative to use (although, on many occasions, there will be a very limited choice). Not all preservatives are active against all micro-organisms, or at all acidities. Therefore, the formulation of the food will often go a long way to determining which preservative to use. How much to use will be determined mainly by how much is actually needed (assuming that it is not above the legal maximum): there is no point in adding levels that are excess to requirements, as this merely wastes money.

How do preservatives work?

Different preservatives work in different ways. Sulphur dioxide and the sulphites, bisulphites and metabisulphites inhibit many reactions within microbial cells such as energy production, protein biosynthesis, DNA replication and membrane function, although with such a range of targets for inhibition, it is not clearly understood which are the most important. It may be a cumulative effect that results in antimicrobial action, or different sites of inhibition may predominate in different situations. Sulphites are especially useful in inhibiting yeasts and moulds in low pH and low water activity products, and Enterobacteriaceae and other Gram-negative bacteria in higher pH and high water activity products (Russell and Gould, 1991). Sorbate and benzoate are also known to inhibit a number of enzymes in micro-organisms. Those containing sulphydryl (-SH) groups may be specific targets.

The antibiotic nisin appears to work primarily by depolarising the microbial cell membrane. In effect, it causes transient pores to appear in the cytoplasmic membrane, which results in loss of small components such as potassium ions and amino acids.

The effectiveness of any antimicrobial can be very dependent on the nature of the food, such as its relative fat and water content, and the pH. Some compounds may be less soluble in the matrix of fatty foods than they are in water-based foods. Acidity is also particularly important, as it can change the chemical nature of the antimicrobial, and one form may be more active than another. The sulphites and bisulphites provide a good example of this (see Box 17).

Antioxidants

In addition to preservatives, antioxidants are widely used to prevent chemical deterioration of foods; this includes the rancidity caused by oxidation of fats, and the browning of cut vegetables caused by the formation of polymers after the action of the enzyme polyphenol oxidase (see Box 18).

Sulphur dioxide and the sulphites are very good at preventing water-based oxidation reactions, as is ascorbic acid (vitamin C). It is interesting to note that, when it is used as a vitamin, ascorbic acid has to be declared as such in food labelling, but when it is used as an antioxidant (i.e. its purpose is to have an effect on the food, not on the consumer!), it must be declared as an antioxidant, giving the chemical name or 'E' number.

Box 17 - Sulphite species equilibrium affected by acidity

Because of the unusual physical chemistry and atomic properties of sulphur, in particular the energy levels of the electrons in the outer orbitals of the sulphur atom, it is able to combine with oxygen in a number of ratios to form a wide range of 'oxoanions' (Russell and Gould, 1991). Those believed to be of relevance in antimicrobial action are sulphite (SO_3^{2-}), bisulphite (HSO_3^-) and metabisulphite ($S_2O_5^{2-}$). The relative levels of these anions in solution depends on the acidity of the solution.

When dissolved in water, gaseous sulphur dioxide forms what is known as sulphurous acid - H_2SO_3; however, in reality it is believed to form a physical structure called a clathrate, in which the sulphur dioxide is trapped inside a shell of about 7 water molecules without any actual chemical bond formation. At dilute concentrations in acid foods (i.e. those with relatively high hydrogen ion concentrations), a mixture of 'sulphurous acid' and bisulphite exists (the pK value is 1.9: this is the pH value at which the mixture is in a 50:50 equilibrium).

$$H_2SO_3 \Longleftrightarrow H^+ + HSO_3^-$$

In neutral and slightly alkaline (low hydrogen ion concentration) foods, a mixture of bisulphite and sulphite anions exists (pK value is 7.2).

$$HSO_3^- \Longleftrightarrow H^+ + SO_3^{2-}$$

In more concentrated conditions, the bisulphite anion can also condense and dehydrate (i.e. give up a water molecule) to give the metabisulphite anion:

$$2HSO_3^- \Longleftrightarrow S_2O_5^{2-} + H_2O$$

This is also an equilibrium reaction that depends on the prevailing pH - as the pH rises (and hydrogen ion concentration falls), there is more dissociation and the reaction moves to the right (i.e. there is more metabisuphite formation).

The metabisulphite anion is particularly important because it is the form that is most commonly added to food as a preservative. However, sulphite, bisulphite and metabisulphite, as well as sulphur dioxide, all exhibit preservative action.

However, reducing the oxidative rancidity of fats (see Box 19) requires an antioxidant that is more lipophilic (fat-loving). The tocopherols (vitamin E), gallates, butylated hydroxyanisole (BHA) and butylated hydroxytoluene (BHT) are the main lipophilic antioxidants permitted in UK and EU legislation. BHT is mainly limited to commercially used fats and oils; BHA and the gallates can also be used on dehydrated foods and powders. As with preservatives, these antioxidants are only permitted in foods at levels for which there is a proven technological and economic need, and with due regard to any adverse effects on health.

Antioxidants work by mopping up the oxidising substances in and around food (e.g. oxygen itself); in the process they become oxidised themselves, but they help to prevent the other more susceptible components in the food from being oxidised. This is analogous to what happens in biological systems; the main difference is that, in the latter, the antioxidant can be reduced to its original form by 'passing on' the oxidising potential and become active again. In food systems, this usually does not happen, so that the antioxidant will eventually be used up.

Box 18 - Preventing browning in fruit and vegetables

The enzyme polyphenol oxidase catalyses the conversion of monophenols in the fruit and vegetable to ortho-quinones in the presence of oxygen. These then polymerise to form brown polyphenolic pigments. In the cells of the uncut fruit or vegetable, the enzyme and the monophenols are physically separated from each other, but the act of cutting breaks down the cell barriers, the products mix and browning quickly occurs. Browning will also occur in damaged tissues, where enzyme and substrate come into contact with each other: the damage may be caused by physical damage (e.g. bruising of apples) or by chilling or freezing damage (as occurs in bananas).

This browning is a significant factor in limiting the shelf-life of fruit and vegetables, and antioxidants such as bisulphite and ascorbate (vitamin C) cn be used to retard it. They act by mopping up the oxidising potential of the environment, reducing the rate of oxidation of monophenols to ortho-quinones; as a consequence, the subsequent polymerisation to brown pigments is also slowed.

Box 19 - Rancidity of fats

Rancidity is the sensory perception of a change in the nature of the lipids (fats) in a food. There are several types of rancidity, the two main types being oxidative rancidity and hydrolytic rancidity. In hydrolytic rancidity, the triglycerides (in which three fatty acid molecules are joined (esterified) to a glycerol molecule) are broken down to release free fatty acids. In oxidative rancidity, the fatty acid chains themselves are oxidized (typically a double bond in the chain is broken and oxygen is incorporated). Neither change is harmful to health *per se*, but both can result in the perception of an off-flavour. In minor cases, the off-flavour is or can be masked, e.g. by the incorporation of highly flavoured ingredients such as spices into the final product, and the problem is circumvented, but this is not always possible or desirable, and preventative measures are required. Palm oil and coconut oil both have a high content of lauric acid-containing triglycerides, which give a distinctive soapy taste when hydrolysed (Ranken *et al.* 1997).

Cooking oils in general are prone to various chemical changes, some of which result in rancidity (they also become less efficient in the frying process). One way of delaying this is to include permitted antioxidants (such as butylated hydroxyanisole, butylated hydroxytoluene and the gallates) in the product, but other non-chemical methods can also be effective. These include: not overheating the oil, keeping it dry (i.e. drying products before frying where possible) and removing debris from the oil that would otherwise act as a catalyst of some of the chemical reactions involved.

Fatty fish such as herring and mackerel contain a high proportion of unsaturated fatty acids and are prone to oxidative rancidity development. The rate of oxidation is dependent on temperature, but still occurs in the frozen fish. Antioxidants have proved to only have a limited effect, and preventing the access of oxygen to the fish, in combination with low temperature, is the best way to slow the development of rancidity. Glazing, vacuum packaging and coating with batter are three possible routes; preventing dehydration also helps (Ranken *et al.* 1997).

Reference:

Ranken, M.D., Kill, R.C. and Baker, C.G.J. (1997) Food Industries Manual. 24th edition. Blackie Academic and Professional.

Box 20 - Labelling of chemical preservatives and antioxidants

Under UK legislation, which is derived from an EU-wide Directive, preservatives, antioxidants and any other permitted additives that are added to a food product to have a technological effect must be appropriately labelled. Either the approved chemical name or its 'E' number (or both) must be used, along with prefix 'preservative' or 'antioxidant' as appropriate. In some cases, the additive may have more than one potential effect; in these situations, the primary reason for using the additive should be considered when labelling. Sulphur dioxide and its derivatives are a good example of this: in some situations they are added for their antioxidant activity (e.g. to prevent browning in peeled potatoes), whereas in others they are added for their antimicrobial activity.

Other additives

As well as antimicrobial and antioxidant preservatives, there are other chemicals that can be added to food to play a role in its preservation: emulsifiers and stabilisers, in particular, help to preserve the physical nature of the food (e.g. the emulsion of an ice cream or the aerated texture of a mousse). Ingredients such as eggs contain lecithin (a phospholipid), which is a natural emulsifier, and which itself is used widely in the food industry. Many others are also used. As with antioxidants and antimicrobial preservatives, the use of these additives is closely controlled: there must be a technological and economic need, and this has to be balanced against any health implications. EU legislation which has been incorporated into UK law specifies which additives can be included in which products and at what levels (Anon, 1995b).

5.2 Curing

Strictly speaking, curing actually means saving or preserving, and processes include sun drying, smoking and dry salting, which are historical processes still in use today. Parma ham, for example, is preserved using relatively low levels of salt and a controlled drying regime. However, the term curing is now generally applied to the traditional process that relies on the combination of salt, nitrate and nitrite to effect chemical preservation of the food, usually meat, but also to a lesser extent fish and cheese.

In the curing of meat, the salt has preservative and flavour effects, while nitrite also has preservative effects and contributes to the characteristic colour of these products. Salt concentrations above 4% in the water phase of the meat inhibit most meat spoilage micro-organisms, mainly by lowering water activity (i.e. the availability of the water for microbial growth). Nitrite further restricts the range of organisms that are able to grow. Some organisms such as the micrococci and lactococci will still be able to grow and will eventually spoil the product, although this can be further delayed if the product is chilled or partially dried (thus increasing the relative salt content). Cured meat products can also be canned with a relatively mild heat treatment, and be shelf stable - the level of salt and nitrite preventing the outgrowth of spores of *Clostridium botulinum,* while the heating process is sufficient to destroy the spoilage organisms that could otherwise grow under these conditions.

Typical cured products are bacon, ham and gammon, and there are a number of variations on the curing technique. Ranken *et al.* (1997) describe the traditional Wiltshire or tank curing of bacon, which relies on the development of a microbiological community to maintain the preservative factors in the process. Although the preservation is clearly due to specific chemicals, it is fundamentally controlled by microbial action. Initially, brine is injected into the pig carcass, with the additional use of dry salt for preservation in the pockets where bones (such as the shoulder blade) have been removed. Immersion follows in a curing brine typically containing 24-25% salt, 0.5% nitrate and 0.1% nitrite. The curing brine containing the established bacterial population is used from one batch to another, being 'topped up' between batches, and is a characteristic deep red colour due to the high concentration of protein that accumulates. Extra salt may be added to counteract the effect of juices seeping out of the meat. Clearly, in such a process, close control is required to ensure that a safe and stable product is produced. Historically this would have required a high degree of empirical expertise and experience. Nowadays the scientific fundamentals are well understood. One of the key factors is the control of nitrate conversion to nitrite. A near-saturated salt solution facilitates the establishment of micrococci and lactobacilli, which convert nitrate to nitrite, and also helps to suppress spoilage organisms and the development of off-flavours. Low temperatures also help in this. The bacterial population also needs to be balanced so that a supply of nitrate is maintained.

One variation on the above is dry curing where, after the initial injection, dry curing salts are added to the meat instead of it being immersed in brine. This tends to result in a drier, saltier product, with increased shelf-life. Historically, this was often carried out in close proximity to tank curing, and the microbial community of the latter would become established in the dry cure area. More recent developments have seen the replacement of the microbial community with sufficient directly added nitrite along with salt to effect preservation. This provides a simple example of how knowledge of a partially microbiological preservation process has allowed it to be converted to a solely chemical process. Although the subtle complexities of microbial metabolism will be lost, and the product sensory characteristics may be slightly altered, the preservation factors will be largely unchanged.

More information on the preservative effects of salts is given in Section 4 on drying and water activity.

5.3 Smoking

This is another traditional process that partially relies on chemicals to effect preservation of the product. Meat smoking derives from the practice of hanging meats in the chimney or fireplace to dry out. This had a variety of effects: the meat was partially dried, which itself assisted with preservation, but chemicals in the smoke also had direct preservative and antioxidant effects, as well as imparting a characteristic flavour on the product. Cheese can also be preserved by smoking, but it is smoked fish that is the most widely produced smoked product in the UK.

Smoking of fish is typically a combination treatment, and involves prior salting of the fish. The effect of smoking on the quality and shelf-life of the fish will depend on a variety of factors: prior preparation of the raw material, type of smoking and characteristics of the smoke itself, and the length of time of the smoking operation. There are basically three types of smoking: cold smoking (where the temperature does not exceed 30°C); hot smoking in conditions that cause thermal denaturation of the proteins (although the actual temperature inside the fish can vary considerably, say from 50°C to 80°C); and hot smoke-drying. However, many local variations exist, as is typical of a process that has evolved at a local community level. These will take into account the type of fish available locally and the normal weather encountered, amongst other things (Doe, 1998).

In a typical operation, following salting, evaporation of water due to drying in air/smoke occurs. Moisture loss in lean fish is significantly higher than in fatty fish of the same species, and this is one of the variables that has to be allowed for in controlling the process to yield a palatable end-product with a suitable shelf-life. Hardwood is most commonly used in traditional smoking operations. Wood smoke consists of thousands of chemicals, of which about 350 have been identified including: carboxylic acids, alcohols, esters, lactones, aldehydes, ketones, ethers, furans, aromatic and aliphatic hydrocarbons and nitrogenous compounds. These individual chemicals have different roles to play in the smoking operation. Some are responsible for the smoky colour and flavour of the final product, whilst others may play no role, contribute undesirable flavours, or indeed be hazardous to health. Many of the components may have direct antimicrobial activity, although some work has suggested that by far the most significant factor is the drying of the fish and the reduction in water activity. The compounds with most antimicrobial activity are probably the carboxylic acids, phenols and formaldehyde, with vegetative forms being more susceptible than spores, and bacteria being more affected than fungi (Doe, 1998). In such a complex preservation system, assessing with any degree of certainty which factors are responsible for most preservation is virtually impossible. The relative levels of each of these antimicrobial components in smoke will depend on a number of factors (e.g. type of wood, temperature of smoke generation), and many of the compounds may show a synergistic effect with each other and with water activity reduction and heating effects. Many smoke components also show antioxidant activity, especially the phenols, and this is also highly significant in product preservation.

In modern smoking techniques, the degree of drying and of smoke deposition and cooking are usually controlled separately, thanks to a more complete understanding of the complex effects that smoking has on flavour, colour, fish preservation and generation of toxic components. Some of the things that can be done to minimise the deposition of toxic components include:

- minimising smoking time
- stopping tar from being deposited on the fish
- keeping the smoke generation temperature to between 350°C and 400°C (this helps prevent the formation of polycyclic aromatic hydrocarbons)
- chilling the smoke prior to the smoking operation to separate out high-boiling components.

5.4 Pickling

This commonly refers to the preservation of foods in acetic acid or vinegar, although the term can be applied to salt preservation. Most food poisoning bacteria are unable to grow at the acidity levels (pH 4) attained during the pickling process, although a much higher degree of acidity (pH 1.5-2.3) is needed to prevent growth of yeasts and moulds (see Table 8).

A number of vegetables are pickled in vinegar in the UK, such as beetroot, gherkins and cucumbers, onions and cabbage, as well as walnuts and eggs. Mixed pickles are also available. In some products, the raw or cooked material is simply immersed in vinegar to effect preservation, but in others, additional processes such as pasteurization are required to produce a palatable and safe end-product (e.g. pickled beetroot is peeled and usually pasteurized in vinegar after cooking).

The acids used in the pickling process exert their effects by interfering with microbial metabolism. All cells need to keep their internal pH within a fairly narrow range. They do this by actively transporting in or out excess hydrogen ions and other ions to keep things in balance. However, if the outside hydrogen ion concentration becomes too great (i.e. it becomes very acidic), then it becomes

Box 21 - pH and acidity

Acidity is related to the concentration of hydrogen ions in a solution. Pure water (H_2O) dissociates to a small degree to hydrogen ions and hydroxyl ions, thus:

$$H_2O \Longleftrightarrow H^+ + OH^-$$

However, the equilibrium of the reaction is greatly towards the water molecule, and the concentration of both H+ and OH- is 10^{-7} moles/litre. The pH value is the negative log of the hydrogen ion concentration. Thus, the pH of pure water is 7. Adding acids such as hydrochloric acid (HCl, which readily dissociates to H+ and Cl-) to pure water increases the concentration of H+ and so lowers the pH value. Adding alkalis such as sodium hydroxide (NaOH, which readily dissociates to Na+ and OH-) to water increases the OH- concentration of the solution, which depresses the H+ concentration (the product of the H+ and OH- concentration is always 10^{-14}).

increasingly difficult for the cell to 'keep up'. In a simple example with vinegar, acetic acid in the surrounding environment dissociates to a certain extent, thus:

$$CH_3COOH \Longleftrightarrow CH_3COO^- + H^+$$

Acetic acid, which is the active component of vinegar used in pickling processes, is a 'weak' acid, which means that the balance of the reaction is to the left (i.e. the undissociated acid predominates over the dissociated acetate anion and hydrogen ions). The cell transports the undissociated acid into the cell, where it partially dissociates, so leading to a build up of hydrogen ions and a lowering of pH.

Table 8 - pH thresholds for growth of some micro-organisms

Micro-organism	Minimum pH
Yeasts and fungi	1.5-2.0
Lactobacilli	3.0
Some bacilli and clostridia	4.0
Escherichia coli	4.4
Clostridium botulinum	4.5
Salmonella typhi	4.5

As with salting, curing and the reduction of water activity in foods, the scientific principles underpinning traditional pickling have been adapted such that some form of acid preservation is used in a wide range of products. Primary food materials tend to be either about neutral pH, low-acid or high-acid (foods are rarely alkaline). Therefore, it is the response of potential pathogens and spoilage organisms to degrees of acidity that is crucial in food preservation. For example, in mayonnaises and other sauces, the acidity of the product, in combination with its water activity and other factors, will determine whether the product is stable at room temperature or whether it will require refrigeration for long-term stability. It will also influence whether or not a heat treatment (pasteurization) step is required. It may be that a combination of acidity, water activity and heat treatment results in an ambient stable product, but that it will require refrigeration after opening. However, the acidity may be high enough, in conjunction with other factors, for the product to be stable at room temperature, even after opening. It very much depends on the formulation.

The threshold for acidic stability is around pH 4.0-4.5. Most fruits are below this threshold (see Box 22), whereas vegetables, fish and meat are above. In canned products, this is highly significant: *Clostridium botulinum* spores are unable to germinate and grow below about pH 4.3. Therefore, many fruit products can be canned with a heat process that does not destroy them, whereas meat and vegetable products require a complete 'botulinum cook'.

One of the most important points to consider when measuring pH in a product, is the homogeneity of the product. Whereas the average pH might be below 4.3, and therefore indicate a significant preserving effect, there may well be 'pockets' within the food that are significantly higher and which would allow microbial growth to occur. This is a potential problem with products such as tomato paste. As with other acid fruit products, this can be canned without a full sterilization process, but it is less acid than many other fruits (see Box 22) and there is an increased chance of parts of the product being above pH4.3. Therefore it is important to ensure that the product is homogenous to avoid the possibility of *C. botulinum* growth and toxin production. With formulated products, such as meat in tomato sauce, even though the sauce portion may be high-acid, the meat portion will be low acid (pH above 4.3) and a full sterilisation process will be required for such canned products.

Box 22 - Typical pH values of certain foods

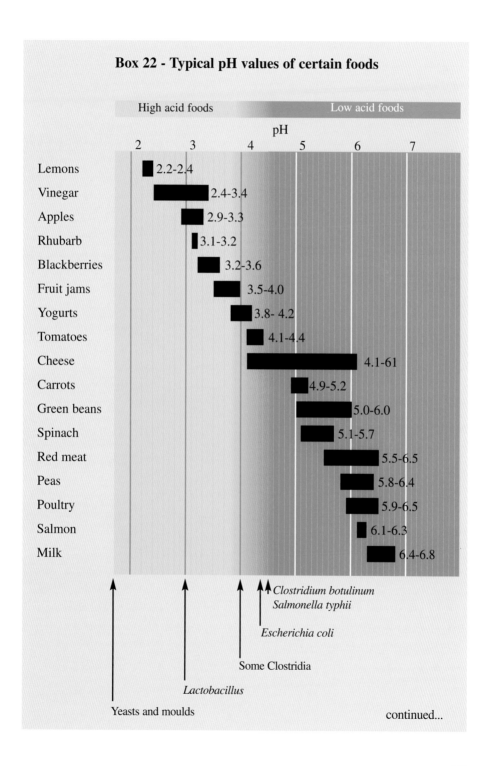

High acid foods   Low acid foods

pH

| | 2 | 3 | 4 | 5 | 6 | 7 |

Lemons 2.2-2.4

Vinegar 2.4-3.4

Apples 2.9-3.3

Rhubarb 3.1-3.2

Blackberries 3.2-3.6

Fruit jams 3.5-4.0

Yogurts 3.8- 4.2

Tomatoes 4.1-4.4

Cheese 4.1-61

Carrots 4.9-5.2

Green beans 5.0-6.0

Spinach 5.1-5.7

Red meat 5.5-6.5

Peas 5.8-6.4

Poultry 5.9-6.5

Salmon 6.1-6.3

Milk 6.4-6.8

Clostridium botulinum
Salmonella typhii

Escherichia coli

Some Clostridia

Lactobacillus

Yeasts and moulds   continued...

Micro-organisms vary in their sensitivity to pH. Yeasts and moulds are relatively acid tolerant, for example, and can grow below pH 2 whereas for most bacteria the threshold is pH 4 or above. High acid foods are products in which all the components have a pH below 3.7 and consequently the significant spoilage organisms for these products are yeasts and moulds and, occasionally, *Lactobacillus* species. Low acid foods with a pH above 4.5, are prone to spoilage with a wider range of micro-organisms. If the product is a mixture of high acid and low acid components, pockets of higher pH can exist where a wider range of micro-organisms can grow, so in these products the pH cannot be relied upon to provide an adequate preservation effect.

Reference:

Safefood Library (1999) CCFRA.

6. MICROBIAL PRESERVATION SYSTEMS

Fermenting foods is one of the oldest ways of preservation - it evolved particularly in warm climates where it was not possible to use natural freezing or chilling facilities to extend the shelf-life of food. Major areas of microbially preserved (fermented) foods are dairy products (such as yoghurts and cheeses), meat products and vegetable products. Bread and alcoholic beverages also use fermentation in their producton, although not primarily for preservation purposes. By adding specific 'starter' organisms, the growth of pathogens and spoilage organisms can be inhibited: the starter culture 'out-competes' them. Non-microbial spoilage - such as biochemical rotting and drying out - can also be side-stepped. In many cases, the food itself is changed so that it is intrinsically less susceptible to both microbial and chemical spoilage. For example, the microbial cultures may result in food either having a much reduced water content or being much more acidic, both of which are methods of chemical preservation discussed above. Also, as discussed in Section 5, chemical preservation by curing can be partially achieved through microbiological systems.

Unlike most other forms of preservation, microbial preservation often involves taking an 'unstable' starting material and changing it to a stable end-product that is very different in terms of eating quality (e.g. cheeses are quite dissimilar from the starting material - milk - from which they are derived).

6.1 Fermented dairy products

The variety of products that can be produced from just one starting material - milk - is quite staggering. There are many hundreds of different yoghurt and cheese products produced commercially throughout the world. Subtly different cultures and conditions result in products with significantly different sensory characteristics. Although originally developed as a way of preserving and using the nutritional value of milk, the end products have become so successful, that many non-dairy analogues have been developed to mimic the characteristics of cheese and yoghurt.

Cheese manufacture

Cheese manufacture probably evolved in the 'Fertile Crescent' between the Tigris and the Euphrates rivers about 8000 years ago (Fox, 1993). The domestication of plants and animals probably led to the 'discovery' of the nutritive value of milk. It would also have been quickly discovered that milk is not stable in warm climates. However, an initial derivative was more stable. This was formed by the action of lactic acid bacteria, which utilise the lactose in the milk, converting it to lactic acid. This lowers the pH of the milk (i.e. makes it more acidic), which causes the caseins (milk proteins) to precipitate. This results in a gel forming, which entraps the fats present in the milk, yielding an edible product that lasts a little longer than the original milk.

This acid milk gel, when broken or cut, yields two components: curds and whey. The whey will not keep and was probably drunk immediately, but the curds could be stored for future use. The shelf-life of the curds could be significantly extended by either dehydration or salting.

Alternatives to the lactic acid bacteria acidification system were quickly noted. Proteolytic enzymes from various sources can modify casein, causing it to coagulate under certain circumstances. The effects of rennet from the stomachs of young animals would have been quickly noticed - animal stomachs were widely used as storage vessels, and when used for the storage of milk they would have been found to yield a product similar to that produced by the lactic acid bacteria. However, it would have been noticed that the resulting product was more versatile - rennet curds can be used to produce low-moisture (extended shelf-life) cheese curd without hardening.

It has been estimated that there are now about 500 varieties of cheese produced commercially. Despite the fact that they evolved as a result of the need to find a longer-lasting product based on milk, they are a generally not 'stable' products, and the changes that occur within the stored cheese can be manipulated to yield the variety of products that exist today. In modern cheese manufacture, the acidification is initiated by adding a defined culture of lactic acid bacteria to pasteurized or raw milk. Typical culture organisms are *Lactococcus lactis*, *Streptococcus salvarius* and *Lactobacillus* species. Acidification is sometimes used to initiate the casein coagulation step, but this is more usually effected by the use of proteinases. Animal rennet has progressively been replaced as the proteinase by fungal enzymes and by chymosin from bacterial strains genetically manipulated to carry the calf chymosin gene.

The final stage in the initial cheese manufacturing process is salting. After this, the maturing or ripening stage begins. This may take up to two years, and is the major source of the variation in cheese characteristics that exists. The complex set of biochemical changes that occur during ripening depend on:

- the initial coagulant type
- indigenous enzymes in the milk - such as proteases and lipases - especially in raw milk
- starter culture bacteria and their associated enzymic activity
- secondary microflora - which may be micro-organisms that survived the initial pasteurization process or which have gained access to the curd after pasteurization; or cultures which have been deliberately added after the initial manufacturing stage, such as *Penicillium roqueforti* in blue cheese

Storage time, temperature and humidity are also important. Thus, the cheese manufacturing process is a complex set of micro-organism-catalysed events, combined with acidification and salting preservation steps, designed to promote beneficial changes while inhibiting deleterious ones.

Yoghurt production follows a broadly similar set of steps - but subtle differences result in a significantly different end product, usually with a much shorter storage life. Tamime and Robinson (1999) describe in detail the many aspects of yoghurt production.

Box 23 - Sauerkraut production

Sauerkraut consists of fermented shredded cabbage and is probably the most well-known fermented vegetable product of European origin. In fact, it is based on an initial dry salting process to provide the conditions for the desired microbiological flora to develop. The shredded cabbage is evenly mixed with 2-2.5% salt - this results in tissue fluid from the cabbage being drawn out by osmosis to form a brine. A 3-6 week fermentation period then follows, in which there is a sequential development of *Leuconostoc mesenteroides*, *Lactobacillus plantarum* and *Lactobacillus brevis*. The final pH of the food is around 3.5-3.7 and this yields a product that is stable for many months (Ranken *et al.* 1997).

6.2 Fermented meat products

Fermented meat products include a range of traditional sausages such as pepperoni, salami and chorizo, as well as some ham products. As with fermented dairy products, meat fermentation processes involve a complex set of microbial changes which have various effects on acidity, water activity and redox potential. These chemical changes, combined with the evolution of a 'benevolent' microflora to out-compete spoilage and pathogenic organisms, help to produce an end-product that is more stable than the raw meat starting material. Originally, the production of such preserved meats would have been an empirical development, but as with the development of dairy products, as more has been understood about the changes occurring during the fermentation, starter cultures have been adapted and fermentation conditions modified to produce stable products of acceptable quality.

The section on hurdle technology (Section 10, Box 29) documents an example of how a fermented salami product is rendered microbiologically stable by an interwoven series of chemical and microbiological changes.

7. MODIFIED ATMOSPHERE PACKAGING

Modified atmosphere packaging (MAP) can be thought of as a form of chemical preservation, in that the gas surrounding the product in a package is not air, but an artificial mixture, usually of carbon dioxide, nitrogen and oxygen, although other gases such as argon have been investigated to a certain extent. It is a technique that is being increasingly used to prolong the shelf-life of fresh foods such as meat, fish and cut fruit, as well as various bakery products, snack foods and other dried foods. The basic idea is to replace the air in a package with a gas composition that will retard the deterioration of the food (e.g. 5% oxygen and 95% nitrogen). In most cases, it will be microbial growth that is inhibited, but in dried foods and fatty foods, the onset of rancidity and other chemical changes can be delayed. The exact composition of the gas used will depend entirely on the nature of the food and the biological process that limits shelf-life and needs to be controlled (see Air Products, 1995; Day, 1992; Day, 2001), and will also vary with the type of plastic used in the packaging of the product. Different packaging materials will have different properties, especially regarding gas and water vapour permeability (see Hutton, 2003 for more details on the types of packaging material used and its characteristics).

For example, in MAP of vegetable pizza samples at 8°C, a 70/30 mixture of nitrogen and carbon dioxide under polyamide/polyethylene film resulted in lower microbial counts in the product than after packaging in air under microperforated polyvinyl chloride, or after packaging in 80/20 oxygen/nitrogen under oriented polypropylene. However, as Figure 2 demonstrates, when noodle stir fry was the product, and *Clostridium botulinum* was the organism of interest, the 80/20 oxygen/nitrogen mixture was more effective. See Box 25 for more information on high-oxygen MAP.

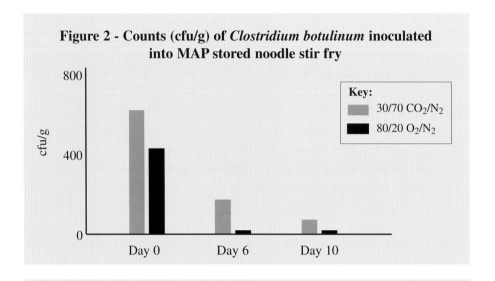

Figure 2 - Counts (cfu/g) of *Clostridium botulinum* inoculated into MAP stored noodle stir fry

Key:
- 30/70 CO_2/N_2
- 80/20 O_2/N_2

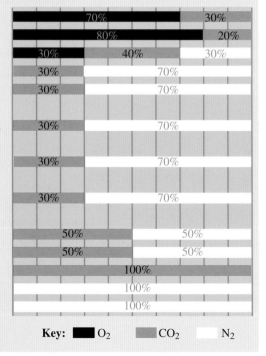

Figure 3 - Typical gas mixtures used in MAP of retail products
(for further information see Air Products, 1995)

	O_2	CO_2	N_2
Raw red meat	70%		30%
Raw offal	80%		20%
Raw, white fish and other seafood	30%	40%	30%
Raw poultry and game		30%	70%
Cooked, cured and processed meat products		30%	70%
Cooked, cured and processed fish and seafood products		30%	70%
Cooked, cured and processed poultry products		30%	70%
Ready meals and other cook-chill products		30%	70%
Fresh pasta products		50%	50%
Bakery products		50%	50%
Dairy products		100%	
Dried foods			100%
Liquid foods and drinks			100%

Key: O_2 CO_2 N_2

A variation on MAP is equilibrium modified atmosphere (EMA) packaging, which is particularly useful for products like fruit and vegetables that continue to respire, using up oxygen and producing carbon dioxide. Developments in plastics technology have produced semi-permeable materials that allow some of the unwanted gases to escape from the pack while retaining those that are beneficial, thus prolonging the product's shelf-life.

MAP is generally used in combination with refrigeration to extend the shelf-life of fresh, perishable foods, but it is also useful for dry, ambient-stable products such as crisps, poppadoms and other snack foods.

Characteristics of gases used in MAP

Carbon dioxide has bacteriostatic and fungistatic properties and retards the growth of most moulds and aerobic bacteria, making it useful in the MAP of moist foods. However, some micro-organisms are not inhibited: lactic acid bacteria grow better in the presence of carbon dioxide, combined with low oxygen levels. Yeasts are also less susceptible to carbon dioxide than moulds and aerobic bacteria.

The inhibitory effect of carbon dioxide is increased at lower temperatures (in addition to the beneficial effect of the lower temperature *per se*). This is because of its enhanced solubility in water (present in the food) to form carbonic acid, which has antimicrobial properties.

The absorption of carbon dioxide into the food is highly dependent on the water and fat content of the food. Excess absorption may cause the package to collapse, through the creation of a partial vacuum; apart from the unappealing nature of this, it might also result in problems with pack integrity, which may result in the seals being broken and the modified atmosphere being lost. The water-holding capacity of meat and seafood products may be reduced, resulting in an unsightly drip, and some products, such as dairy foods, may be tainted. On a more general note, fresh fruit and vegetables, which continue to respire after packaging, may suffer physiological damage in the presence of elevated carbon dioxide levels (Day, 1992).

Box 24 - Other MAP gases

Other gases used experimentally and, to a limited extent, commercially to increase the shelf-life of foods include carbon monoxide, ozone, ethylene oxide, nitrous oxide, helium, neon, argon, propylene oxide, ethanol vapour, hydrogen, sulphur dioxide and chlorine. Many of these have important health or other safety concerns associated with them. For example, carbon monoxide has been shown to be very effective at maintaining the colour of red meats and inhibiting the decay of plant tissue, but it is toxic and is not a permitted packaging gas in the European Union. However, it has been used commercially in Norway for fresh meats (Brydon, 2002).

However, argon, which is chemically inert like nitrogen, may have wider applicability. Its atomic size is similar to the molecular size of gaseous oxygen and it is denser and more soluble in water than either nitrogen or oxygen. It may thus have biochemical activity, being more effective at displacing molecular oxygen from cellular sites and enzyme receptors, consequently slowing down oxidative deterioration reactions. There are over 200 products packed using argon in the UK, including nuts, crisps, pizzas, meats and drinks (Brydon, 2002).

Both argon and nitrous oxide, which are permitted for use as packaging gases in the European Union, are also thought to sensitise micro-organisms to antimicrobial agents. This may be mediated through alteration of the fluidity of components of microbial cell membranes. A French company, Air Liquide, has filed a number of patents concerning the activity of argon and nitrous oxide (N_2O) in MAP. One of these claims that both are capable of extending shelf-life by inhibiting fungal growth, reducing ethylene emissions in fresh produce, and slowing down sensory quality deterioration.

Further reading:

Day, B.P.F. (2001) Fresh prepared produce: GMP for high oxygen MAP and non-sulphite dipping. CCFRA Guideline No. 31.

Fath, D. and Soudain, P. (1992) Method for the preservation of fresh vegetables. US Patent No. 5,128,160

Brydon, L. (2002) Developments in MAP and active packaging. Proceedings of Minimal Processing Conference, Sardinia.

Box 25 - High oxygen MAP

Recently there has been much research into the use of high-oxygen MAP for shelf-life extension of fresh prepared products (e.g. chopped fruits and vegetables) and combination products such as sandwiches, pizzas and chilled stir-fry ready meals. The initial thrust was in the prolonged preservation of items that were still respiring and therefore consuming oxygen and producing carbon dioxide and water. These present a special problem for food preservation over extended periods, as the relative changes in gas and moisture content will affect respiration rate and ultimately the quality of the food in quite a complex manner. One way of circumventing this is by using packaging material with specific permeability characteristics, depending on how rapidly the items are respiring. High-oxygen MAP (i.e. 70-100% oxygen) uses a different approach. The high level of oxygen delays the problem of oxygen depletion and subsequent growth of anaerobic micro-organisms. It was also found to inhibit aerobic micro-organisms, which are adapted to oxygen levels of around 21% (i.e. atmospheric levels). It is hypothesised that reactive oxygen species damage vital cellular macromolecules and thereby inhibit microbial growth when oxidative stresses overwhelm cellular antioxidant protection systems.

Biochemical deterioration also appears to be inhibited. The main problem with cut fruits and vegetables is discoloration, caused by oxidation of natural phenolic constituents to colourless quinones, which subsequently polymerise to coloured melanins. This is initiated by the action of the enzyme polyphenol oxidase, and it seems as if high levels of oxygen directly or indirectly inhibit the activity of the enzyme.

Research was subsequently carried out to evaluate the effect of high-oxygen MAP on the shelf-life of stir-fry ready meals, pizzas and sandwiches, all of which have both respiring and non-respiring components. High-oxygen MAP was found to be more effective than nitrogen/carbon dioxide MAP for extending the shelf-life of both pizzas and ready meals, but no such benefit was found with sandwiches, where the critical factors determining shelf-life (bread staling and moisture migration), are not affected by MAP.

Further reading:

Day, B.P.F. (2001) Fresh prepared produce: GMP for high oxygen MAP and non-sulphite dipping. CCFRA Guideline No. 31.

In most MAP applications, oxygen levels are kept as low as possible, in order to inhibit the growth of aerobic pathogens and spoilage organisms, and to reduce the rate of oxidative deterioration of foods. However, there are exceptions: oxygen is needed for fruit and vegetable respiration, and for colour retention in red meats - in both cases, lack of oxygen will significantly reduce the quality shelf-life of the product. Residual levels of oxygen will also prevent the growth of anaerobic pathogens such as *Clostridium botulinum.* (However, when designing a food preservation strategy and determining or assigning shelf-life, it is important to be aware of the danger of residual (aerobic) organisms using up the remaining oxygen, and thus creating conditions suitable for the anaerobes.) The biochemical and antimicrobial effects of high levels of oxygen have also been researched recently (see Box 25).

Nitrogen is effectively inert and has a low solubility in both water and fat. In MAP, nitrogen is used primarily to displace oxygen in order to retard aerobic spoilage and oxidative deterioration. It also acts as a filler gas to prevent pack collapse.

The choice of the gas combination to be used will depend on the nature of the food, and of the micro-organisms associated with that food which will limit the shelf-life. It is also important to bear in mind that preventing the growth of one type of micro-organism may allow another group to proliferate. The most notable example of this is in the creation of anaerobic environments, which may allow the growth of *Clostridium botulinum,* an organism that can only grow in the absence of oxygen, but which produces a highly potent neurotoxin when it does grow.

The effectiveness of any modified atmosphere system will also be dependent on the type of packaging material used to enclose the gas, including its permeability to gases and water vapour. This is discussed in more detail in Hutton (2003).

8. VACUUM PACKAGING

Vacuum packaging takes some of the theories of MAP one stage further. In general, MA packed foods tend to be processed and then packaged with the introduction of the modified atmosphere; vacuum packed foods can be produced in a similar way, i.e. processed and then packaged with the introduction of a vacuum. However, they can also be packed under vacuum and then processed (i.e. cooked) within the package. The latter form of processing mainly originated in France and is termed 'cuisson sous vide' (cooked under vacuum).

The main reason for vacuum packaging is to remove all oxygen from the packaged food. The main effects of this are to eliminate the problem of oxidative rancidity of fatty foods, and the growth of spoilage or pathogenic bacteria that require oxygen to metabolise and survive. The potential downside of vacuum packaging is the growth of anaerobic pathogenic organisms such as *Clostridium botulinum*. Thus, whilst vacuum packaging eliminates some 'problems' so that food can be stored for prolonged periods of time, it usually needs to be carried out in combination with other preservation mechanisms in order to gain maximum benefit in terms of quality and extended shelf-life. As *C. botulinum* is the most significant pathogen in this context, it serves as a good example of what needs to be considered when producing vacuum-packaged foods. Growth of most strains of *C. botulinum* is inhibited at refrigeration temperatures; however, psychrotrophic (cold-growing) strains do exist. This means that chilled, vacuum packed (or 'sous vide') products need to have extra preservation factors to prevent its growth. An important point to make is that products which are not chilled at all are analogous to canned products - to be stable at room temperature, they must either have received a full sterilization process or have an inherent preservative factor that prevents growth of mesophilic *C. botulinum* (e.g. have a high salt content, low pH, or be cured or dried). Products that have not received a sufficiently high level of heat process can only be safely stored for extended periods of time under strictly controlled conditions. Fortunately, there are a number of controls that can be used to achieve this.

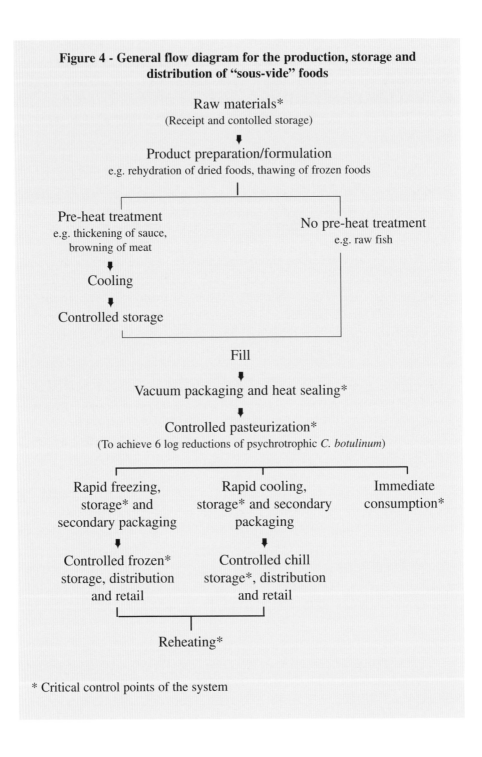

Figure 4 - General flow diagram for the production, storage and distribution of "sous-vide" foods

Raw materials*
(Receipt and contolled storage)

Product preparation/formulation
e.g. rehydration of dried foods, thawing of frozen foods

Pre-heat treatment
e.g. thickening of sauce,
browning of meat

No pre-heat treatment
e.g. raw fish

Cooling

Controlled storage

Fill

Vacuum packaging and heat sealing*

Controlled pasteurization*
(To achieve 6 log reductions of psychrotrophic *C. botulinum*)

Rapid freezing,
storage* and
secondary packaging

Rapid cooling,
storage* and secondary
packaging

Immediate
consumption*

Controlled frozen*
storage, distribution
and retail

Controlled chill
storage*, distribution
and retail

Reheating*

* Critical control points of the system

Storage at 3°C or below is effective at preventing the growth of *Clostridium botulinum*. At this temperature, it is likely that other factors will determine the shelf-life of the product. However, it is very important to ensure that this low temperature is maintained, and this is only practically feasible in predistribution storage of the product. Thus, it is only useful in extending the storage life of the product before distribution.

Storage at 8°C or less can be used to prevent the growth of *C. botulinum* for up to 10 days, without any other preservation mechanism. This is the baseline from which all chilled vacuum-packaged products should start when a shelf-life is being established. To extend the life beyond 10 days, one or more of several alternative preservation mechanisms has to be employed: a pH of less than 5.0, a salt concentration of 3.5% w/v in the aqueous phase of the food, or a water activity of less than 0.97. In all cases, these minimum levels have to be achieved throughout the food in order to achieve an extended shelf-life. Alternatively, a heat process equivalent to 90°C for 10 minutes can be used. Depending on the extent of these extra preservative factors, and other shelf-life limiting factors within the food, the product can be safely kept for more than 10 days.

Figure 4 gives a general flow diagram for the production, storage and distribution of "sous-vide" foods (Betts, 1992). As the stages involved have such a major impact on the safety of the final product and its achievable shelf-life, strict control of the production process is vital. The most important steps, the 'critical control points' are indicated.

9. NOVEL TECHNIQUES

Food technologists are continually looking for new ways to produce food with enhanced flavour and nutritional characteristics. Traditional thermal preservation processes do tend to reduce the vitamin content of food and can affect its flavour. The development of processes which are as effective as traditional thermal systems in reducing or eliminating micro-organisms and effecting preservation, but which have less of a deleterious effect on the nutritional constituents of the food, are being actively developed. Some of these are based on existing techniques, while others use completely different forms of energy to reduce microbial numbers. For example, ohmic heating is a novel way of heating a food by passing an electric current through it. There have also been developments in drying and freezing techniques to enhance product quality.

Several novel methods of food preservation are discussed in detail in Leadley *et al.* (2003). Power ultrasound, food irradiation, pulsed light, pulsed electric fields microwaves, high-pressure processing and electric field and magnetic field systems are all being actively investigated.

Since May 1997, the introduction of a completely 'novel' process has been controlled in the EU by European Parliament and Council Regulation No 258/97 (Leadley *et al*, 2003). This sets out a statutory pre-market approval system for novel foods, including foods produced by novel processes throughout the European Union and requires companies who wish to market a novel food to submit an application to the Competent Authority in the Member State where they first intend to market the product. In the UK this is the Food Standards Agency, which is advised by the Advisory Committee on Novel Foods and Processes (see Box 26).

High pressure processing

The idea of preserving foods and drinks by using very high pressures to reduce or eliminate micro-organisms and enzymes was originally thought of in the 1890s, but it was not until the 1970s that Japanese food companies started to explore its potential. Pressures of several hundred times that of atmospheric pressure are used, in cycles of short bursts, to kill unwanted micro-organisms. Jams were the first products to be produced in this way in Japan, and the process has also been investigated and developed in Europe and the USA. Among the products that have been produced commercially using high-pressure processing are oysters and various dips in the USA, and fruit juices and similar products in Japan, the USA and Europe. High-pressure-treated tapas produced in Spain is now available in the shops in the UK. For more information see Leadley *et al.* (2003). An example of the use of these extremely high pressures is given in Box 28.

Ohmic heating

This is a thermal process, but instead of applying external heat to a product to have a sterilization effect, an electric current is applied directly to the food; the electrical resistance of the food to the current causes it to heat it up. The technique does not rely on transfer of heat from outside of a product into the product (e.g. as in a conventional oven) and the heating of the product is relatively rapid (around 1°C per second) and occurs simultaneously throughout the food. Thus, much shorter heating times can be applied than would otherwise be possible, and so the product will potentially maintain more of its nutritional and flavour characteristics. Frankfurter style sausages are particularly suitable for this type of process, as are viscous liquids containing solid particles.

Irradiation

This technique has seen much wider applications in the USA than in the UK, where 'public opinion' has effectively sidelined it. In addition to killing bacterial pathogens, such as *Salmonella* on poultry, it is especially effective at destroying the spoilage micro-organisms present on fresh fruit such as strawberries and thus

Box 26 - Role of ACNFP

In the UK, before any completely novel food, novel ingredient or food produced by a novel process can be marketed, it has to be considered by the Advisory Committee on Novel Foods and Processes (ACNFP), which advises the Food Standards Agency of its suitability and safety. The ACNFP is an independent body of scientific experts which advises government of the safety of these novel products. Although most of the applications it receives are for novel ingredients, it also looks at the products resulting from novel processes. For example, in 1998, the Committee generally agreed with the opinion of French Competent Authority on the suitability as a human food of fruit preparations pasteurized using a high-pressure treatment process.

Article 5 of the Novel Food Regulation (EC) 258/97 allows the notification of the intention to market a novel food or food produced via a novel process, based on a scientific opinion of a particular Competent Authority (in the UK, the Food Standards Agency as advised by the ACNFP). Under this procedure, the company concerned can notify the European Commission of their intention to market a novel food, which they consider to be as safe as a conventional counterpart. Notifications must demonstrate that in terms of nutritional value, metabolism, intended use and level of undesirable substances the novel food is substantially equivalent to an existing safe conventional counterpart. This notification gives authorities the opportunity to challenge if they disagree with the company's judgement.

Box 27 - Active packaging

Although not novel in the sense defined by ACNFP (see Box 26), active packaging is a new and innovative approach which involves using the package to help effect preservation. The term active packaging refers to the incorporation of certain additives into packaging film or containers with the aim of maintaining and extending product quality and shelf-life. Packaging may be termed active when it performs some desired role in food preservation other than providing an inert barrier to external conditions. The development of a whole range of systems, some of which may have application in both new and existing food products, is fairly new.

continued....

Active packaging includes additives or "freshness enhancers" that are capable of:

- scavenging oxygen
- adsorbing carbon dioxide, moisture, ethylene and/or flavour/odour taints
- releasing ethanol, sorbates, antioxidants and/or other preservatives
- maintaining temperature control.

Oxygen scavengers are the most commercially important sub-category of active packaging. They can help maintain food product quality and thus increase shelf-life by decreasing food metabolism, reducing oxidative rancidity, inhibiting undesirable oxidation of labile pigments and vitamins, controlling enzymic discoloration and inhibiting the growth of aerobic micro-organisms. The most well-known oxygen scavengers take the form of small sachets containing various iron based powders with an assortment of catalysts. These chemical systems often react with water released from the food to produce a reactive hydrated metallic reducing agent, which scavenges oxygen within the food package and irreversibly converts it to a stable oxide. The iron powder is separated from the food by keeping it in a small sachet which is highly permeable to oxygen and is also labelled 'Do not eat'. The main advantage of using such oxygen scavengers is that they are capable of reducing oxygen levels to less than 0.01%, which is much lower that the typical 0.3-3.0% residual oxygen levels achievable by modified atmosphere packaging (MAP). Oxygen scavengers can be used alone or in combination with MAP or vacuum packaging.

Ethylene is a natural plant hormone which accelerates the senescence (aging) of horticultural products such as fruit, vegetables and flowers. From a food preservation point of view, slowing down the aging or ripening process in fruit and vegetables will potentially aid in preserving the product and extending its shelf-life. This is particularly the case with products that lose their green colour during ripening, such as tomatoes and bananas. Effective systems utilise potassium permanganate immobilised on an inert mineral substrate such as alumina or silica gel. Potassium permanganate oxidises ethylene to acetate and ethanol and in the process changes colour from purple to brown, thus indicating its remaining ethylene scavenging capacity. More information on active packaging systems can be found in Hutton (2003).

Reference:

Hutton, T. (2003) Food packaging: an introduction. CCFRA Key Topics in Food Science and Technology No. 7.

Box 28 - Extremely high pressure - a novel method for improving flavour

In 2001, Orchard House Foods introduced 'Extremely High Pressure' (EHP) treated orange juice to the UK. It claimed that a 'low temperature, high pressure system' that kills bacteria naturally present in fresh foods, whilst significantly extending shelf-life, is ideally suited to fresh orange juice. Because the EHP treated juice is not heated, as it is with pasteurization, the flavour profile in the orange juice is not affected in the same way.

In 'blind' taste tests conducted by Orchard House to compare fresh, EHP and pasteurized orange juice, EHP juice emerged virtually identical to fresh, whilst pasteurized juice was felt to taste more processed, less sweet and less tangy and juicy. 88% of consumers stated that EHP treated juice tasted 'fresh' compared to just 56% who thought the same of pasteurized juice. From a visual perspective, nine out of ten consumers also thought that EHP juice looked natural and appetising, compared to 73% who felt this was true of pasteurized juice.

The company claimed that EHP treated orange juice retained all the natural characteristics of fresh orange juice whilst extending shelf-life from the normal nine days to up to 21 days.

Orchard House was the first UK company to install this particular piece of technology, but the system had already proved successful in the US and mainland Europe where it had been used to extend the shelf-life of short shelf-life products such as guacamole and fresh apple juice. It has also been used successfully to process oysters without loss of flavour.

markedly extending their shelf-life. It can also be used to prevent sprouting in potatoes. Both its biggest advantage and biggest disadvantage are that it has so little effect on the food itself that it is very difficult to tell if the food has been irradiated. This is a big plus for product quality, but raises problems as to how to ensure that consumers are informed of the nature of the food they are buying. Much work was done in the 1990s to develop ways of detecting irradiated ingredients.

In practice, irradiation is an expensive process to carry out and so commercial applications are generally restricted to low volume/high value products. It also has some technical limitations, in that it is not suitable for products that are high in fat, as it can lead to the generation of off-flavours. The only commercial food products that are currently licensed for irradiation in the UK are dried herbs and spices, which are notoriously difficult to decontaminate by other techniques, without markedly reducing flavour.

In the UK and the EU, there is a requirement to label food that has been irradiated or that contains irradiated ingredients.

10. COMBINING PRESERVATION TECHNIQUES

There has been a consumer drive in recent years for preserved foods to look and taste more like their fresh counterparts. This might mean reducing the degree of preservation (and hence often the shelf-life) of the product. However, by using several different preservation mechanisms in combination, and by reducing the degree of each individual preservation step, such products can be produced.

Although the term 'hurdle technology' is relatively new, its application is not: pasteurization of foods is often combined with some other preservation step to achieve an extended shelf-life. This might be an extrinsic preservation mechanism, such as chilling, or may be intrinsic to the food itself (e.g. the acidity of vinegar-based sauces). In some cases, combining acidity with chilled storage may allow the pasteurization step to be drastically reduced or even eliminated.

Traditional preservation techniques often involve a number preservation hurdles. Hot smoking is a combination of mild heat processing, reduction of water activity and direct chemical preservation, and is often combined with a salting phase as well. Fermentation procedures mediate their activity through a change in acidity as well as competitive microbiological mechanisms, and curing involves changes in acidity and water activity, and direct chemical effects. The hurdle combinations of acidity, water activity, temperature, heat processing, 'natural' chemical preservatives and storage gas composition, whether inherent or developing within the food itself, or imposed from the outside, are almost limitless - some examples are given in Box 29.

Box 29 - The hurdle technique - modern examples

The novel application of hurdle technology in the preservation of foods began in the 1970s in Germany, for the preservation of meat products such as fermented salami sausages. In effect, a process that had been operated for many years was analysed and the individual hurdle preservation steps were identified, enabling
continued....

them to be controlled better. In salami fermentation, nitrite and salt levels are important in the early stages of ripening, as they inhibit many bacteria in the initial product, such as pseudomonads and other oxidative bacteria, which will rapidly multiply and spoil uncured meat. Other bacteria will proliferate and use up the available oxygen, reducing the redox potential of the product and therefore inhibiting oxidative bacteria, but allowing the growth of lactic acid bacteria. These in turn metabolise the sugars added to the product, resulting in an increase in acidity (a decrease in pH), which is a significant preservative hurdle. Although the preservative effects of redox potential and pH decrease subsequently (both rise in long-ripened salami as nitrite and lactic acid bacteria levels decrease), water activity (a_w) decreases as the product ripens, and this becomes the major preservative factor. These changes control the growth of pathogens and indigenous spoilage organisms in the product. Undesirable mould growth on the surface of the sausage can be prevented by smoking or the addition of chemical preservative (such as sorbate), or by using desirable mould starter cultures (Leistner and Gould, 2002).

The above is an example of controlling a dynamic process to enable preservation hurdles to be maximised. Combinations of hurdles can also be applied in more 'static' situations. For example, carefully controlling ingoing salt and nitrite levels in canned ham can allow the product to be given a reduced heat process (i.e. the product does not need a full sterilization because the salt and nitrite will prevent the outgrowth of surviving spores of *Clostridium botulinum*).

There has been a major consumer push over recent years for minimally processed fruit and vegetables. Washed, peeled/cut or trimmed produce can be packaged in a modified atmosphere and stored chilled, with the two hurdles providing an extended shelf-life over that achievable for the unprocessed product. Vacuum-packed foods can be given a pasteurization process and stored chilled for up to 10 days. To achieve a longer shelf-life, additional preservation hurdles need to be present. These may be salt, acidity or specific preservatives. More recent developments have investigated the use of natural preservatives from herbs and spices, organic acids and high-pressure treatment (Leistner and Gould, 2002).

Further reading:

Leistner, L. and Gould, G.W. (2002) Hurdle technologies: combination treatments for food stability, safety and quality. Kluwer Academic/Plenum

11. SHELF-LIFE OF PRESERVED FOODS

The length of time for which any particular food can be kept will depend on the nature of the food itself, and the preservation treatments to which it has been subjected. It is up to the manufacturer of the food to determine and assign the shelf-life of the food they produce, keeping in mind the requirements of national and international regulations. In the UK the Food Safety Act requires that a food must be both of the nature, substance and quality expected by the consumer, and also not injurious to health for whatever reason. Thus, it is illegal to sell food which has deteriorated during storage so as to become a health hazard, or if its quality has deteriorated beyond that which would be normally acceptable. Food manufacturers go to great lengths to ensure that this does not happen.

Determination of shelf-life

In determining the shelf-life of a particular food, the first pre-requisite is to know what particular characteristic of the food is going to be the limiting factor of its shelf-life. Determining that biscuits have a microbiological shelf-life of several years is of no interest, if they are going to become soft and not of edible quality after 3 months. Sometimes, there may be more than one factor that needs to be assessed. For example, the shelf-life of chilled, raw meat pies might be limited by micro-organism levels, but there may also be a quality issue with rancidity development, or with water or fat migration from the meat into pastry, causing the latter to become soggy.

The food processor or manufacturer should have enough knowledge of his product to be able to determine what factors are going to limit the product's shelf-life, and the approximate time for which the product will remain fit for eating (i.e. days, months or years). To fix a more precise time requires one or more of the following:

- extended trials (i.e. storing the product under the required conditions and testing it for the critical parameters at various times)
- accelerated shelf-life trials (usually keeping the product at a higher temperature than would normally be the case and relating the shelf-life achieved under these conditions to those that might be achieved under 'normal' conditions of storage)
- a thorough comparison of the shelf-life achieved with similar products already being produced
- the use of microbial growth models for predicting when microbial growth might reach a critical level and render the product unacceptable

In practice, all of these have limitations and combinations of all might need to be looked at for some products, in order to reach a safe and sensible decision.

It is beyond the scope of this book to look in detail at shelf-life evaluations, but it is worth pointing out the advantages, disadvantages and limitations of each method. One limitation with all trials, particularly of microbiologically unstable foods, is the assumption that the consumer will follow the storage instructions given on the package. Allowance must always be made for a reasonable degree of product abuse.

Extended trials

Keeping the product under the proposed storage conditions and monitoring chemical, microbiological and/or organoleptic changes that occur over time is the only way of accurately determining shelf-life and is well suited to short shelf-life products (i.e. those with shelf lives measured in days or a few weeks). However, if the shelf-life is likely to be measured in several months or years, it can be impractical to have to wait that long before being able to market the product. It may be possible to compare the change in the food characteristics being measured with those of an existing product for which the shelf-life is already known, to see if the change in characteristic is faster or slower. Care must be taken when doing this to make sure that the factors which limit shelf-life do not change (and the wrong parameters are monitored), and that the existing and new product formulations are very similar. This 'method' can only ever be indicative.

Accelerated shelf-life trials

These trials usually use storage at a higher temperature than normal (or some other 'exaggerated' storage condition) to speed up the deterioration of the food, and are another potential way of estimating the shelf-lives of long-life foods. What is not always appreciated is that different food quality and safety parameters are affected to different degrees by, for example, increases in temperature. Therefore, the approximate response of the food or its ingredients at ambient and elevated temperatures needs to be known in order for the trials to have any meaning. What also needs to be known is the food quality factor that limits the shelf-life (i.e. the factor that 'goes off' first), so that the trials can be designed to measure this factor. This may change under different conditions. The use of storage conditions that are significantly different to those which are likely to occur in reality in the storage of the product may also initiate changes that would not normally be seen. Great care therefore is needed in using accelerated shelf-life trials, which can only be indicative.

However, for many products, there is not a simple relationship between what happens under normal and extreme conditions - especially in products where physical deterioration is the main problem (e.g. crisp/crunchy biscuits becoming soft).

Challenge testing

Challenge testing is designed to answer the following specific question:

"Will the product formulation and storage conditions control growth of pathogens during the designated (or proposed) shelf-life of the product, if they were present in the food?"

It involves deliberately adding pathogenic micro-organisms (or safe alternatives with similar growth characteristics) to the food in a laboratory situation to see if they can survive and grow under typical product storage conditions (Betts *et al*, 2004; CCFRA, 1987). Obviously, in some foods, experience and/or mathematical models will show that microbial growth is not possible, and that microbial challenge testing is unlikely to yield any useful information (see Figure 5).

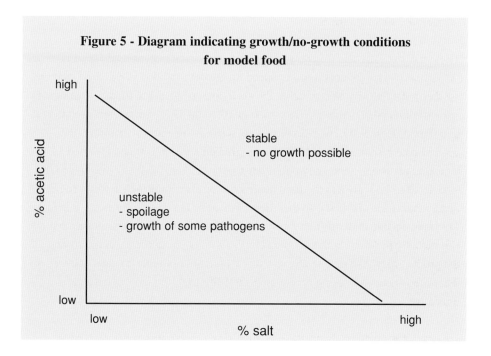

Figure 5 - Diagram indicating growth/no-growth conditions for model food

It is important to use existing knowledge and mathematical growth models, if appropriate, in order to 'home in' on areas of potential concern.

In other situations, the likelihood of the presence of pathogens in the product after processing may be so small as to make challenge testing of no practical value. It may be possible to demonstrate that a pathogen can grow in a product, but the challenge test does not indicate the likelihood of it being there in the first place. This is an important distinction which highlights the need to understand the product and the pathogen, and emphasises that these tests need to be carefully planned and expert knowledge is required for the interpretation of results.

Box 30 - Determining shelf-life in chilled foods

Deciding for how long a particular chilled food will remain fit for consumption is fraught with difficulties. In the majority of cases where the shelf-life is limited by microbial growth, chilling will only delay the onset of growth and/or slow its rate. If there are micro-organisms present and they are able to grow, eventually they will. Even with the best mathematical models, accurately estimating the degree of slowing is difficult - micro-organisms are notoriously unpredictable! In many other chilled foods, non-microbial changes will be the determining factors.

There are many factors that may affect the timing and rate of microbial growth, such as water activity, salt or sugar content, acidity, and added preservatives. The type of processing the product has previously received and how it is packaged will also be of relevance, as this will influence the likelihood or otherwise of the presence of pathogens and/or spoilage organisms (and the level of contamination). To compound these factors, the food may well have a very different shelf-life after it has been opened to that before. This applies not only to foods where microbial growth is likely to be the limiting factor, but also to fatty foods (e.g. paté) where oxidative rancidity is likely to occur.

In practice, when developing a chilled product, shelf-life evaluation is done at three stages: at kitchen/pilot plant scale; in full-scale factory trials; and in full-scale on-going production (Betts *et al*, 2004). This is because the actual formulation may well change as the product and process is tested - and minor changes in product formulation can have a significant influence on shelf-life. Although the shelf-life of many chilled foods is related to microbial activity, it is often the result of the activity (rather than the microbe itself) that can be measured to determine shelf-life (e.g. volatile amine production in fish and meat, and headspace gas analysis in yoghurts). In other products, non-microbial changes need to be measured, e.g. sogginess of pastry, loss of crispness of vegetables, and rancidity development in fatty foods.

Reference:

Betts, G.D., Brown, H.M. and Everis, L.K. (2004) Evaluation of product shelf-life for chilled foods. CCFRA Guideline 46.

Challenge tests are of most value when the organisms used represent those that might realistically be present in the product, either from the ingredients or via post-process contamination. The levels added in the tests also need to be realistic - adding large numbers of *Listeria*, say, when contamination levels, if they occur, are almost certain to be low, is of limited value.

Challenge tests, unlike the extended and accelerated storage trials described above, will not give an estimation of shelf-life, but they will indicate whether the shelf-life proposed is inherently safe.

Mathematical models

There has been much work done in recent years into the development of mathematical models to predict the timing and rate of growth of micro-organisms in foodstuffs. The general idea is that, by knowing certain characteristics of the food, such as its acidity, salt and water content, storage temperature, and perhaps the presence of specific preservatives, a prediction can be made of how long the food will remain microbiologically acceptable. The development of these models is not straightforward. Individual micro-organisms, even different strains within the same species, can have very different growth characteristics, and this growth will be affected by the food matrix, the general availability of nutrients and the physiological state of the organism itself, the number of organisms initially present, and the likely competition for nutrients from other organisms. Despite this, significant progress has been made.

The first point to consider is what degree of growth is unacceptable and signifies the end of the product's shelf-life. With pathogens, any growth at all might be unacceptable, and models need to predict the 'time to growth' of the organisms initially present. With spoilage organisms, presence and slight growth may not be a problem: spoilage organisms are not a safety issue *per se*, and if the food itself shows no sign of spoilage in terms of taste, odour or visual changes, then it may be perfectly acceptable. In these cases, the mathematical models need to predict the time it will take until spoilage organisms build up to an unacceptable level, or begin to have a deleterious effect on the food. A third scenario is to predict whether particular organisms can actually survive under a particular set of conditions. In this case, the microbial growth model becomes a microbial death model.

Having decided what, microbiologically, is going to limit the shelf-life of the food of interest, a series of experiments has to be devised which cover the range of the conditions likely to be encountered in the food. These might be: up to 5% salt, pH4.0-6.0, storage at 2-10°C, and presence of 2-4% acetic acid. It is not necessary to investigate growth under every conceivable combination of conditions, but it is necessary to include the extreme conditions; i.e. one growth trial must be with 0% salt, pH6, 2% acetic acid, and storage at 10°C (the most 'lenient' of conditions), with another at 5% salt, pH4, 4% acetic acid and storage at 2°C (the most stringent of conditions). It is important to be aware that the model will not be valid outside of the conditions chosen; i.e. it can not be used to predict growth at 12°C, or pH7 or in the absence of acetic acid.

In some circumstances (e.g. in spoilage growth models) it may be possible to devise the experiments using natural foods, without any addition of micro-organisms. In this case the models would predict the growth of the microflora inherently present in the food. When doing this, it is important to take into account the likely variability in the natural microflora - both the numbers and types of micro-organisms present.

In other cases, e.g. in the prediction of survival and growth of pathogens in foods, it may be that the pathogen is not usually present in the food of interest (e.g. *Escherichia coli* in minced meat). In these cases, it is necessary to inoculate the food with the pathogen of interest, or with a 'safe' alternative that has similar growth characteristics.

Having gathered the data from the chosen series of experiments, it is possible to make broad predictions of shelf-life for any given set of circumstances within the boundaries of the conditions investigated.

Box 31 - Shelf-life labelling of foods

Virtually all processed and packaged foods have to be labelled with an indication of their 'minimum durability', i.e. their shelf-life.

The form in which these shelf-life date marks are given depends upon the nature of the food involved, the likely duration of its shelf-life and the limiting factor for that shelf-life. Short shelf-life products that are likely to become hazardous due to the action of micro-organisms must be marked "use by", together with any storage conditions that should be observed (often "keep refrigerated" or something similar). The date specified should include the date and month - for example, 'Use by 30 June'.

Other products that require a date mark are labelled "best before".

Products with a shelf-life of less than 3 months are marked "Best before" with the date, month and, optionally, year. In this case, both 'best before 30 June' and 'best before 30 June 2004' would be acceptable.

Products with a shelf-life of between 3 and 18 months can be marked "Best before end" with the date alternatively specified as just month and year - for example 'Best before end June 2004' would be equally as acceptable as 'Best before 30 June 2004.

Products with a shelf-life of more than 18 months can be marked "Best before end" with the further alternative of just specifying the year. In this case the date could be given as just 'Best before end 2004', or as 'Best before end December 2004' or as 'Best before 31 December 2004'.

12. CONCLUSIONS

Food has been preserved since mankind first developed culinary skills, initially by chilling and freezing in cold climates, and salting and fermenting in warmer climates. Drying was also an early preservation mechanism, and the development of international food trading boosted this form of preservation (removing water from food also results in considerable weight saving, which is still very important for food transport economics).

On an industrial scale, canning was the first major development towards long-term mass preservation of a wide variety of primary products. Until then, the only mainstream products traded that had shelf-lives measured in months or years were dried foods. In the domestic situation, canned and dried foods reigned supreme until the commercialisation of frozen foods in the mid 20th century. This was followed relatively soon by the development of the chilled food industry, boosted by the rapid spread of domestic refrigerators and the desire for a wider range of products with improved eating quality that could be stored for more than a few days, but did not necessarily have to be kept for months.

Modern developments and improvements to these basic preservation techniques, such as aseptic processing, and the addition of novel techniques, such as modified atmosphere packaging, to be used in conjunction with existing techniques, have resulted in a previously unimaginable choice of medium to long shelf-life products available in the shops. The competition between these preservation techniques has itself led to improvements in quality as each has tried to achieve the advantages of the others, e.g. a longer shelf-life in chilled foods, improved texture in frozen foods and a more 'fresh-like' flavour in canned and dried foods.

13. REFERENCES

Those references in bold were of particular use to the author in compiling this overview and the reader may find them of interest if wanting to delve further into the subject of food preservation.

Air Products (1995) The Freshline guide to modified atmosphere packaging.

Air Products (1999) Modified Atmosphere Packaging Gas Selector. CD-RoM

Anon (1995a). Food Safety (Temperature Control) Regulations 1995. SI 2200

Anon (1995b) The Miscellaneous Food Additives Regulations 1995. SI 3187

Anon (1992) The Workplace (Health, Safety and Welfare) Regulations 1992. SI 3004

Anon (1990) The Quick-frozen Foodstuffs Regulations 1990. SI No. 2615

Anon (2003) The Jam and Similar Products Regulations 2003. SI No. 3120

Bell, C. and Kyriakides, A. (2000) *Clostridium botulinum.* **A practical approach to the organism and its control in foods. Blackwell Science**

Betts, G. D. (1992) The microbiological safety of sous-vide processing. CCFRA Technical Manual No. 39

Betts, G.D., Brown, H.M. and Everis, L.K. (2004) Evaluation of product shelf life for chilled foods. CCFRA Guideline No. 46.

Betts, G.D. (1996) Code of practice for the manufacture of vacuum and modified atmosphere packaged chilled foods with particular regard to the risks of botulism. CCFRA Guideline No. 11.

Betts, G.D. and Gaze, J.E. (1992) Food pasteurization treatments. CCFRA Technical Manual No. 27.

British Standard 7915 (1998) Ergonomics of the thermal environment. Guide to design and evaluation of working practices for cold indoor environments

Brown, K.L. (2000) Guidance on achieving reasonable working temperatures and conditions during production of chilled foods. CCFRA Guideline No. 26

Brydon, L. (2002) Developments in MAP and active packaging. Proceedings of Minimal Processing Conference, Sardinia.

CCFRA (1999) Safefood Library

CCFRA (1987) Guidelines for microbiological challenge testing. CCFRA Technical Manual No. 20

Chapman, S. and Baek, I. (2002) Gums and thickeners: a review of food hydrocolloids. CCFRA Review No. 34

Day, B.P.F. (1992). Guidelines for the good manufacturing and handling of modified atmosphere packed food products. CCFRA Technical Manual No. 34.

Day, B.P.F. (2001) Fresh prepared produce: GMP for high oxygen MAP and non-sulphite dipping. CCFRA Guideline No. 31.

Department of Health (1994) Guidelines for the safe production of heat preserved foods.

Doe, P.E. (1998) Fish Drying and Smoking: Production and Quality. Technomic Publishing

Fast, R.B. and Caldwell, E.F. (1990) Breakfast cereals and how they are made. American Association of Cereal Chemists

Fath, D. and Soudain, P. (1992) Method for the preservation of fresh vegetables. US Patent No. 5,128,160

Fox, P.F. (1993) Cheese: Chemistry, Physics and Microbiology. Volume 1. General Aspects. 2nd Edition. Chapman and Hall

Haisman, D.R. and Knight, D.J. (1967) Beta-glucosidase activity in canned plums. Journal of Food Technology 2: 241-248

Hershkovitz, E. and Kanner, J. (1970) The effect of heat treatment on beta-glucosidase activity in canned whole apricots. Journal of Food Technology 5: 197-201

Hutton, T. (2001) Introduction to hygiene in food processing. CCFRA Key Topics in Food Science and Technology No. 4

Hutton, T. (2003) Food packaging: an introduction. CCFRA Key Topics in Food Science and Technology No. 7

Leadley, C.E., Williams, A. and Jones, J.L. (2003) New technologies in food preservation: an introduction. CCFRA Key Topics in Food Science and Technology No.8

Leistner, L. and Gould, G.W. (2002) Hurdle Technologies: Combination Treatments for Food Stability, Safety and Quality. Kluwer Academic/Plenum

Lopez, A. A. (1987) A Complete Course in Canning and Related Processes. Book 1 - Basic Information on Canning. 12th Edition. The Canning Trade Inc.

May, N.S. (1997) Guidelines for establishing heat distribution in batch overpressure retort systems. CCFRA Guideline No. 17

May, N.S. (2002) Analysis of temperature distribution and heat penetration data for in-container sterilization processes. CCFRA Review No. 22

Orchard House Foods (2001) Trade press release of 4th July 2001 on extreme heat processed orange juice. www.ohf.co.uk 12/12/01

Ranken, M.D., Kill, R.C., and Baker, C.G.J. (1997) Food Industries Manual. 24th Edition. Blackie Academic and Professional.

Rose, D. (1986) Good manufacturing practice guidelines for the processing and aseptic packaging of low-acid foods. CCFRA Technical Manual No. 11.

Russell, N.J. and Gould, G.W. (eds) (1991) Food Preservatives. Blackie Academic and Professional

Singleton, P. (1997) Bacteria in Biology, Biotechnology and Medicine. 4th Edition. John Wiley & Sons

Singleton, P. and Sainsbury, D. (1978) Dictionary of Microbiology. John Wiley & Sons

Smout, C. and May, N.S. (1997) Guidelines for Performing Heat Penetration Trials for Establishing Thermal Processes in Batch Retort Systems. CCFRA Guideline No. 16

Stanier, R.Y., Adelberg, E.A. and Ingraham, J.L. (1976) General Microbiology. 4th edition. MacMillan Press p4

Tamime, A.Y. and Robinson, R.K. (1999) Yoghurt: science and Technology. Second Edition. Woodhead Publishing Ltd.

Thorne, S. (1986). The History of Food Preservation. Parthenon Publishing.

Thorpe, R.H. and Barker, P.M. (1985) Hygienic design of post process can handling equipment. CCFRA Technical Manual No. 8

Thorpe, R.H. and Everton, J.R. (1968) Post-process sanitation in canneries. CCFRA Technical Manual No. 1

Timperley, A. W. (1997) Hygienic design of liquid handling equipment. CCFRA Technical Manual No 17 Second Edition

Tucker, G. (1995) Guidelines for the use of thermal simulation systems in the chilled food industry. CCFRA Guideline No. 1

Voysey, P.A. (1999) Guidelines for the measurement of water activity and ERH in foods. CCFRA Guideline No. 25

ABOUT CCFRA

The Campden & Chorleywood Food Research Association (CCFRA) is the largest membership-based food and drink research centre in the world. It provides wide-ranging scientific, technical and information services to companies right across the food production chain - from growers and producers, through processors and manufacturers to retailers and caterers. In addition to its 1600 members (drawn from over 60 different countries), CCFRA serves non-member companies, industrial consortia, UK government departments, levy boards and the European Union.

The services provided range from field trials and evaluation of raw materials through product and process development to consumer and market research. There is significant emphasis on food safety (e.g. through HACCP, hygiene and prevention of contamination), food analysis (chemical, microbiological and sensory), factory and laboratory auditing, training, publishing and information provision. To find out more, visit the CCFRA website at www.campden.co.uk